THIS BOOK IS DEDICATED TO my favorite ghost excavating "buddy", my daughter, Melissa Sabol. I will miss her critical reviews of my ideas and fieldwork operations. But I know she will be with me in "spirit", as she begins her college studies in Philadelphia. I'll miss ya Mel!!

THE ANTHRACITE COAL REGION:

THE ARCHAEOLOGY OF ITS HAUNTING PRESENCE

by
John G. Sabol Jr.

authorHOUSE®

AuthorHouse™
1663 Liberty Drive, Suite 200
Bloomington, IN 47403
www.authorhouse.com
Phone: 1-800-839-8640

First published by AuthorHouse   3/17/2008

ISBN: 978-1-4343-6896-6 (sc)

Printed in the United States of America
Bloomington, Indiana

This book is printed on acid-free paper.

Art Work by Melissa Sabol

The author and poet, William Pitt Root, once observed:

> "The landscape you grew up in....constitutes
> a kind of primordial reality for each person.
> You will never find a place more real than
> the one you grew up in, where your eyes
> are new and just opening".

"I come from coal
My bones are filled with marrow black
liquid carbon fills my veins
the legacy of men whose faces never saw the light of day
whose lungs were filled with pockets filled with dust
whose hearts beat out a rhythm
as they dug into the earth's jet core....."
 Maggie Chelland Martin, " I Come From Coal"
  (from, Coalseam: Poems from the Anthracite Region)

Preamble

Part 1
Coal: The Anthracite's Region <u>Still</u> Untapped Resource

Coal has always had a "dirty" and soiled reputation. It is now an "old fashioned" source of energy and heat. It is (was) largely a domestic product, to be used in "familiar" living spaces, and was relatively "cheap", especially by today's oil prices. Coal does not "conjure-up" or manifest an image-enriching environment. And the perception of a coal miner is a "poor", but hard-working, individual. Is this what the general public (largely comprised of "outsiders") thinks of <u>our</u> anthracite coal region as well? How very little are they informed!

Coal is so mundane that contempt and social distain still lingers. Today, in the 21st.century, children are <u>still</u> told that, if they misbehave or are naughty, they will receive a lump of coal in their Christmas stocking. While oil, and its by-products, are symbols of good fortune (for those who own the rights), coal is viewed as a disappointing, and limiting, material gain. How dark and obscure is that fact!

There <u>is</u> richness, a largely untapped vein, in the history of coal and its extraction, especially here in this "coal region". In this book, coal, and its related mining operations in this anthracite-rich landscape (unique in the world), is a symbol of still undiscovered "treasure". This is the largely unexploited potential of the ethnic heritage that worked in these coal mines. This untapped source of cultural wealth includes immigrants from 26 different European countries who made this region their new home! One aspect of that ethnic heritage will be explored here. These are the haunting dramas and uncertainties that are associated with the mining of the region's anthracite coal. These excavations are the real ghosts of our past, a past I wish to share with everyone (both the living and the ghosts). This story is rich in social content, human emotion, and historic, day to day struggles, of common, ordinary people.

According to Barbara Freese, in her book, Coal: A Human History (2003):

> "Coal is the highly concentrated vestige of extinct life forms that once dominated the planet, life forms that were themselves a critical link in the chain of environmental changes that made the emergence of advanced life possible" (2003:3).

Are the anthracite ghosts we encounter in this region part of that vestige of "extinct" life forms? Or are they something else, perhaps a more advanced (or alternative) form?

There is precedence for this perception of a coal/ghost connection because there is a long and storied history of an intimate relationship between the two. During the early Bronze Age, for example, the earliest inhabitants of southern Wales used coal to cremate their dead (Hatcher 1993:17). Did these people perceive coal as an appropriate "charm" (or symbol) to help escort the dead on their continuing journey, and their manifestation back (in some instances) to the world of the living?

It is hard for us to visualize today, in this "advanced" technological age, a workplace that is darker, more dismal, and dangerous than an early coal mine setting, be it located in the British Isles (where coal was first largely exploited in the West), or Schuylkill County, Pennsylvania (where the American "coal rush" began). According to Freese (2003):

> "The mines had ceilings that could collapse on your head, air that could smother you, poison you, or explode in your face, and water that could rush in and drown you or trap you forever. Coal mining was one of the few occupations in which a person faced a very real risk of death by all four classical elements – earth, air, fire, and water" (2003:47)

Is it no wonder that this vast and varied haunting uncertainty translated well to a vision of a hell on earth? This "hell" (appropriately underground) was made even <u>more</u> frightening "by the miner's dread that the inexplicable disasters that plagued them were due to demons and goblins (and the ghosts) haunting the mines" (Freese 2003:47).

Even the appearance of the miner himself was sometimes a vision of ghostly presence. Stephen Crane, the author of the Civil War drama, <u>The Red Badge of Courage</u>, wrote what he once saw in a mine:

> "After a time we came upon two men....The garments of the men were not more sable than their faces, and when they turned their heads to regard our tromping party, their eyeballs and teeth shone white as bleached bones. It was like the grinning of two skulls there in the shadows. The tiny lamps in their hats made a trembling light that left weirdly shrouded the movement of their limbs and bodies. We might have been confronting terrible specters".

Truth is sometimes stranger than fiction. What Stephen Crane described were indeed ghosts. These were the "ghosts within" of these "living" anthracite miners!

Any unknown (or unexpected) sound in the mines (it was dark after all) was attributed to ghosts, usually of those miners who were killed in a mining accident. Yet, this did not stop the men from telling ghost stories, and playing "tricks" on other workers, "by pretending to be the spirit of someone who had died" (Bartoletti 1996:63). The mine shafts, and the coal veins, were the depository of more than just heating energy. They were the source of ghostly tales, and ghostly energetic manifestations that continue to this day.

There will always be <u>some</u> purpose for coal, even in the now largely empty spaces of the coal industry collapse here. Don't be surprised, for it has <u>always</u> been so with the usefulness of coal:

> "Any substance versatile enough to pierce ears in Neolithic China, accessorize togas in ancient Rome, smoke out snakes in Dark Ages Britian, darken paint in prehistoric Pennsylvania.... probably has still undreamed-of future uses "
> (Freese 2003: 236-237).

That future use, I suggest, is for <u>today's</u> haunting, and ghostly, stories, and their profitable application. Coal has always been associated with one form of exploitation or another. Perhaps, we should add to these haunting dramas, and continue to exploit the coal. This time, however, for positive reasons. A "ghost rush" may yet serve the same economic purpose as the historic "coal rush" once did, for this developmentally-challenged region. But, let's hope that this "ghost boom" (if it is applied) will be longer-lasting than the previous "King

Coal boom", perhaps for a period that could potentially span multiple times and worlds!

So, let us now begin the journey to personally meet and greet these anthracite ghosts.....

Part 2
Archaeology and the Coal Region Ghost

Ghost research, like archaeology, brings the dead back to life. What was once a "corpse" of dead cultural/ethnic remains, is now an awakened individual and/or group drama (and memory). This is a recall, however fragmentary, of a life of work and activity in these coal mining fields. Ghost research, also similar to archaeological research, is working with the remains (manifestations) of the past to enrich history, to recover than which has been lost to contemporary human memory, and to add details to an understanding of (and possible solutions to) individual pasts. These traces of the past connect ghost research to archaeology, and to all kinds of memory practice and recall. This makes all of us, who use memory to remember events and activities from our past, an archaeologist of sorts. We are all archaeologists of ghostly (and haunting) remembrances.

Archaeology, however, is limited in its ability to tell us much details about specific individuals performing mundane, habitual activities, or an individual's role in particular events, especially if they were young and immature individuals (such as "breaker boys"). This is one advantage a ghost investigation potentially has over an archaeological investigation. Through performance-based ghost excavations, individual ghost "signatures" may be found.

It is not enough that a ghost field investigator function simply as a "hunter" and gatherer of anomalous data and information. There is also a story to tell, a drama that is still occurring in the field. This is an individual experience and encounter with the ghost. It should be the goal of the ghost researcher, as an excavator, to unearth this haunting drama.

A ghost investigator (as "archaeologist"), like the ethnic miner before him, digs down and works hard, to bring out these other worlds and sensory materials into the light and warmth of the day. This extraction of buried "treasures" is only accomplished through blood, sweat, and a "ton" of emotional output and effort. Sometimes, the "raising" involves a "lowering": life begets death/death becomes life. To understand this difference, one has to view the manifesting remains of these haunting dramas through an excavator's eyes, not that of a tourist, or casual bystander. These coal (and cold) remains have to be unearthed, carefully taking into consideration their original context, and this is not an "outside" (or "outsider") view. The excavation, itself, is a form of remembering, and, perhaps, in the process, there is an unearthing of a forgotten (and "miner") memory. These archaeological "digs" are true to their historical nature. They are, in the vast majority of instances, a search for the ordinary and mundane. They follow their kinship ties on a direct path back to their source: the popular "image" of coal, and its extractor, the coal miner.

This archaeology of the mundane and habitual collapses and unfolds time, making irrelevant the distinction between a past, present, or future experience and manifestation:

> "Now that he has departed from this strange world a little ahead of me, that means nothing. People like us....know that the distinction between the past, present, and future is only a stubbornly persistent illusion"
>
> Albert Einstein

This coal region ghost archaeology pays more attention to the everyday existence of people, rather than the grand events of history. It honors the minor (and miner) as a major player in local and regional haunting manifestations. The mine, its underground passages, and the supporting surface structures, are the entrances to a different world, the "undiscover'd country" of the Shakespearian ghost.

The unearthing of these coal region ghost memories (as a form of cultural haunting) transforms their perceived banal content into something else. These haunting dramas contain the continued essence of the "black diamonds" of yesterday. In the 21st c., these "black diamonds" serve a more equitable purpose than in the past, both useful and profitable for the many (as opposed to the few), irregardless of one's ancestry or finances.

We must reawaken our senses from the depths and obscurity of the present, and recall a different historical perspective of this frequently excavated landscape. There are many cultural layers to dig through, but this excavation process is far easier than mining the coal that preceded it. This is because these anthracite fields are symmetrical in nature, and thus capable of unfolding, if only we would use our sensitivity as a tool (rather than our analytical minds), to guide us in this unearthing process.

Archaeological sites resonate with activity, which produce an aura of mystery and discovery that continue to surround and envelop them. The same can be said for the coal region structures and features that dot the landscape (breakers, collieries, mines, culm banks, and the ethnic "patches"). This mystery is based on what remains still exist, what use they had in their heyday, and what happened in between.

In this transitional (liminal state), the many haunting uncertainties of this transformation have created both intrigue, and a ghostly presence. Yet, these manifestations are merely the innate symmetry that is contained within these anthracite coal structures.

An excavation is the only means to discover what has (had) happened in this transitional process. There is beauty in the "discovery": the haunting certainties of all that has occurred within, and around, these structures. The excavation process requires the investigator (or interested party) to stand center stage, in that place of uncertainty, and unearth the "dirt" of transition. This can only be achieved by comparing the structure's present appearance (both perceived and physical), to what is known about its past. This comparison is interpreted through performative actions at the site, and is meant to uncover the existence of a continuing presence (or its absence) from that structure's past cultural history.

The present work is meant to open us to the potential reality of what it means to dig around in the dirt, and, in that innocent child-like action, look for fragments (the bits and pieces of innocence) of that anthracite "coal rush" aura. The excavation is a true archaeological poetics, a human sensitivity for the remembrance of forgotten, mundane histories and personal struggles.

This unearthing is a means to perceive phenomena (usually anomalous) in a new way, and is achieved by viewing things from the "inside out". It is a re-discovery of those hidden "black diamonds" that were physically buried in the earth, but arise, and become visible again, in this archaeological excavation action.

The ghosts that are contained in these pages are bodies that have significance (and substance) that go beyond their immediate historical context and physical setting. The exposing of these anthracite coal ghosts unveils them (potentially) to a variety of contemporary cultural traditions. This is hauntingly poetic, since their origination is also a symmetrical blend of multiple ethnicities. The ghost, in this excavated context, has become impervious to the decompositions of the body. By surviving, even today, in these coal region structures, these ghosts become "famous" (rather than infamous and spooky), just by manifesting their past histories in these haunting dramas in the field. And for many of them, these histories are mundane and habitual. This is a just and fitting reward, a contrast to all the daily toils and struggles that were experienced by these ethnic miners.

The archaeology of the anthracite phantom is a true "treasure" (not ghost) hunt, because by finding and communicating with them, a more enriching coal mining experience emerges, and with it, their stories become a significant, useful, and emotional encounter again:

> "It is as if….. the dead should wander for a space mongst aliens
> to the name and race, the secrets of the grave's abyss to tell"
>   John Hughes

Opening Remarks

"What is it that lures boys to haunted houses after dark, Compelling them to fling rocks and run away....What is it that grips a child....to listen to ghost stories....And beg for more and more?"

Jack London

These questions are some of the founding purposes that underlie a quest for these anthracite hauntings. What follows is a series of mini-test "excavations" -a renewable unearthing- of hauntingly true places, events, and activities that one can encounter in the coal region. These excavations take the form of short essays, rather than embodied apparitions. Nevertheless, they are still spooky because, unfolding within these symmetrical historical narratives, one may encounter, even today, the ghosts of the anthracite past.....

Today, more than ever, life is a continual movement from a remembered path to the still (at least for me) unexpected turn in the road. My childhood memories, though now matured into more adult pursuits of cause and effect relations (an economics in steps and expenditure), they have still created various detours. This traveled road is curved because the innocence of that youthful enthusiasm has not been lost in the process. Where I once began, so do I continue, no matter which paths I have chosen to pursue. Loren Eiseley once said,

> "I am not the first man to have lost his way only to find, if not a gate, a mysterious hole....that a child would know at once led to some other dimension....at the approach of age, some men look about them at last and discover the hole....leading to the unforeseen. By then, there is frequently no child companion to lead them safely through."
>
> (1969:196)

I have followed a different path. I still retain, in memory, my own childhood experiences and innocence, using these qualities as a compass to guide me. It is my own, and personal, "ghost within". The spaces these ghosts inhabit are not unique. They live within all of us, although some are buried more deeply than others. From these remembered experiences, I continue to pursue a route through the gate in the woods, past the cemetery, and the old abandoned coal structures, and into the vast subterranean mines of my mind.

And this is why these anthracite hauntings are a continuation of that ghost hunt that began so long ago. The time machine to that past is both internal and ghostly. The only fuel it requires is an open, and imaginative, grasp of possibilities, rather than certainties. It is the unfolding of a childhood past into the contemporary frame of adulthood, one that is still visualized and sensed with that openness of a young, and more innocent, age. The child that is still present in me is the presence of that ghost within, as I continue to haunt myself. Isn't that what a "true" phantom does anyway? The "haunted houses" that are contained within this mind (mine-?) are the remembered experiences of that youth, searching for a conclusion that still has no end.

It is these haunted houses that I use as the perceptual tool for the continued unearthing of these anthracite fields of drama. These houses of the "living dead" are linked by internal roads. Each one of us has these roads imprinted in our own personal mind episodes. Each one of us has the capacity to chart and map these lost fields of drama.

Our haunted houses are engraved with our personal signatures, and are a hierarchy of structures containing the various experiences and activities of living our lives. The "little treasures" of these houses and fields of drama are not mental constructs. They are real foundations for exploring a "night country":

> "There exists for each one of us an oneiric house, a house of dream-memory, that is lost in the shadow of a beyond of the real past"
>
> (Bachelard 1969:15)

These dream houses, and their buried treasure, need to be excavated, and the memories contained within them, need to be unearthed and brought out into the light of day. When they are exposed, a new alternative world unfolds before us. This is the basic premise for continuing the journey, and is the subject of these anthracite hauntings. _

Table of Contents

List of Illustrations

Preface

The Creed of a Ghost Excavator: The Art and Science of Controlled Memory Recall

"I come to the fields and vast palaces of memory, where are the treasuries of innumerable images of all kinds of objects brought in by sense-perception. Hidden there is whatever we think about....and whatever else has been deposited and placed on reserve and has not been swallowed up and buried in oblivion.... some things ....have to be drawn out...until what I want is freed of mist and emerges from its hiding places".

Saint Augustine

A ghost excavator attempts to "clear" the earth and the dust (just like the miners and "breaker boys" of the past) from memory, such that this unearthing clarifies one perception and vision of those past memory histories. However, there is a special method that is involved in this clearing. Societies cannot remember in any other way that through their ancestor's memories. This is "collected memory" (Young 1993: XI). The "collected memory" that follows in this book is the memory of the anthracite coal region as one group, its ghosts, remembers it.

David Bakhurst (1990) claims that,

> "to remember is always to give a reading of the past, a reading which requires linguistic skills derived from the traditions of explanation and story-telling within a culture….this is true even when what is remembered is one's own past experience….(this) image of the past….becomes a phenomenon of consciousness only when clothed with words, and these owe their meaning to social practices of communication" (1990:219).

Ghost excavation is a very personal practice of communicating with these anthracite ghosts. The process of excavation questions the arbitrary distinction between the past and the contemporary world, such that it becomes fruitless to discuss whether or not a particular event of history that is remembered, sensed, or observed corresponds to an actual past reality (does the handprint on the wall in the Jim Thorpe jail reflect a miscarriage of justice or something else?). What really matters are the specific conditions under which these memories (and their recall) are constructed, and the personal and sociocultural implications under which they are held.

In this anthracite region, these specific conditions concern the "clearing" of the forest, and the subsequent mining of coal. The personal and sociocultural implications of large numbers of migrations to this region created the memories of haunting uncertainties that led to the birth of the coal region ghost. The remnants of that landscape exploitation, the mining industry structures, the historical locations of work and strife, the culm banks, and the semi-abandoned towns and patches, are places where people who grew up in this environment still remember. These fragments of memories, like other artifacts, led (and can still lead) people to create a history through the active remembrances of that past as they occur within the sociocultural context in which they continue to live (and have lived). In these lived and living spaces, the ghost excavator does his best to unearth the haunt dramas that still continue to unfold and manifest here.

So, I, as a ghost excavator, am like others who came before me to this anthracite area, and thus am not unique. We all are what Loren Eiseley has called the "librarians of the night", those who search for "whatever rustles in thickets upon solitary walks" (Eiseley 1969:195). We are those who pursue nocturnal shadows that "have come from the dark wood of the past" because "our minds are haunted by night terrors that arise from….private memories" (Eiseley 1969:195). It was, as if inside of us, there is a shape shifting being circulating within, one that embodies a blood flow that is not contemporary, or necessarily 100% human.

We speak of (and work with) a past, one that is capable of unfolding, if merely in fleeting glimpses and fragments. Time is recalled through experiences at particular locations. There is a certain poetics in this, an inner space deep within the vast caverns of the mind. It is our own personal "deep map" of our interior mines. We excavate because that is the only means to unearth the accumulations of memories that are constantly being deposited there. It is an unending, reiterative task, even occurring <u>after </u>physical death. We are what all of you will <u>eventually</u> become. We are all, in the end, the ghost excavators.

The Anthracite Area and Its Ghostly Presence: How to Read This Book

"What is a ghost? Stephen said with tingling energy. One who has faded into impalpability through death, through absence, through change of manners"

James Joyce, <u>Ulysses.</u>

This book is not about traditional stories of haunted houses or graveyards (most graveyards aren't haunted anyway!). Those locations are mere "Hollywood images", produced for mass consumption, and fueled by the gullibility of the viewer. They appeal to the uninformed because their perception of spookiness is easily invoked. They form part of a long tradition of representation, and thus are easily, and without much effort, recalled. You only have to imagine hard enough, and the ghosts <u>will</u> appear. The imagination is a brilliant engineer, easily and with flawless effortlessness, is able to construct the infrastructure that gives "life" to these ghost stories:

"Oh, I went down to Framingham
to sit on a graveyard wall;
"If there be spooks," I said to myself,
"I shall see them, one and all."....
(For) I knew....the secret of raising a ghost.

And the method is this – at least for a miss –
You must sit on a graveyard wall,
And talk of the things you never have seen
And you'll see them, one and all".

Nathalia Crane, "Spooks"

These ghosts of the anthracite area are harder to imagine and see because they are more academic sorts. One needs to think about how they lived their lives to acknowledge the reasons for their presence. Here, in this region, the ghosts are ethnic and cultural hauntings. They are less solid, and require more "fleshing out". Cultural connections need to be explored. Spatial symmetry has to be outlined and analyzed. There is an unfolding of time in these haunt dramas, as experience and place memory are used to transplant the traditional setting of hauntings to other, less noted, locales and half-forgotten events. In place of the haunted house, we have the haunted mine shafts. We don't see these ghosts often. We hear them more. There is also a forest of spooky symbolisms, where changelings exist. These are not ghosts, so much as transformers, changing their guise from one ethnic tradition to that of another.

Yet, these entities remain similar to traditional haunts in many respects, and as resonant influences, they allow some applicability and conformity to other similar ghostly manifestations that occur elsewhere. They still appear to follow a recognizable pattern of haunting manifestational behavior.

The short essays that follow, and the ghosts contained in them, are rich in haunting imagery. They may invoke a chilling sensation, if absorbed too fast or read at length. For best results, one should sit quietly at night – preferably late (though early evening in winter is just as effective). Choose an isolated, dislocated corner of the room. This is hard sometimes, especially with young children, and in these very "un-silent" row homes. But try to make the best of it. Sit for a while, still and quiet. Under these conditions, some wandering spirit (not necessarily a ghost) might drop-in (perhaps a "deceased" relative or a former owner), invited by this sense of anticipation and calm emotional state. Their stay may not be a prolonged one, a fleeting moment perhaps?

Don't expect much of a conversation. Dead silence with these visitations is the norm. But, be prepared for some sensually-producing manifestations! Remember, these are fragmented appearances, as are these essays. But, rest assured, time will be well-spent! Later, and hopefully, daylight will find you in "high spirits", and refreshed from the experience of both the written narratives, and their ghostly manifestations!

May these readings provide you with ample reasons to "dig up" your own memories and experiences in this very haunted (but seldom explored as such) anthracite region. At the same time, one should be aware that these stories can provide a forum for optimism. As filmmaker, Stanley Kubrick once stated: "ghost stories, no matter how dreadful (and I hope this book isn't), still are ultimately optimistic: they promise shape beyond self". In the anthracite region, or in today's world, it is not always easy to be optimistic. So sit back and enjoy these essays, originating from the pen of one who has experienced these ghosts on a very personal level. They contain the memories that last <u>beyond</u> a lifetime!!

Introduction: Some Pre-Excavation Musings

A. The Anthracite Region "Ghost Map"

"Each of us stands at one unique spot in the universe, at one moment in the expanse of time, holding a blank sheet of paper. This is where we begin".
                    Peter Turchi, <u>Maps of the Imagination (2004)</u>

This book is my map for the anthracite region. It is the map of a ghost excavator. This is a map of a geography that is formulated outside of oneself, and is the basis for the exploration of one's inner ghosts: the emotions, social relationships, experiences, and memories of one's own personal geography of hauntings. It is a form of "spirit managing", and involves the excavation of the inner mines that contain the remembrances of particular landscape settings.

It is in this marginal environment, the anthracite region of the post coal-boom era in which I grew-up, that one had an opportunity (and a necessity) to develop strategies of adaptation and survival. To do otherwise, would be to abandon "home" (and hope), and settle into a commonality of existence of work and play routines. Many who have stayed here all their lives have done just that. They continue to live "routinely". That is why change is slow here. Tradition remains dominant. You perform here in a capacity which has always been familiar. These patterns continue to exist today. And this is the haunting nature of the past here. These patterns eventually created and transformed the haunting uncertainties of the past into various forms of ghostly presence, scattered throughout the anthracite region.

Growing-up in the Mahanoy Area (in the center of this anthracite region), I saw these patterns being born, develop, and, more importantly, remembered. These patterns include (the most remembered of them for me):

- The connection that exists between solitary play and imaginary journeys, and the loneliness of spaces where there are few people with which to interact;
- The relationship between a need to escape the routine, and spaces where one can do so, without attracting attention; and
- The symmetry between a landscape and its abandoned structures that are darkened memories from the past, yet are close and capable of a past still being sensed there.

It has been a great privilege for me to have lived (to continue to live) in a town (and surrounding area) in which the dead have not abandoned. This town (Mahanoy City) is still filled to capacity, even though many of its homes and structures (at last count more than 400) are seemingly deserted and unoccupied. The former residents, in some cases, still inhabit these spaces. The ghosts are everywhere here, perhaps even 24/7.

From its very beginning, this anthracite region has seen people migrate to this area from all over, including Native American migrations. This "polyglottery" has helped to create a rich and varied ethnic tradition. The stories of the drama in the field that unfolded (and still unfolds) here reflects the enormous variety of that cultural and ethnic patchwork.

The anthracite region historical saga is all about the lives of common, ordinary people who became ghosts. If you sat across the dinner table from these people while they were still alive, it would never have occurred to you to imagine that, "this is the type of person that inhabits haunted houses". And that is the secret mystery of the ghost within in this anthracite region. It is a "spirit" of adventure and re-discovery. But it came at a steep price:

> "Mining has rarely been carried on except by pioneers and slaves. It has always been colored by the spirit ofdiscovery and adventure".
>
> John L. Lewis, 1925, President of the
> United Mine Workers of America.

The anthracite landscape incorporated both pioneers (the vast immigrant populations) and slaves (those "chained" to this mining industry). Today, they are all ghosts of the "coal rush". The ghost map of their haunted locations is defined by the coal industry structures that still dot the landscape of this once thriving industry!

B. The "Fields" of the Anthracite Region and Their Ghost Excavation

The fields of drama of the anthracite region are not so much a physical entity situated within defined physical boundaries, as they are areas of performed influences. These areas of performed influences are the "home" of the various haunting phenomenon, and they lay at the very heart of the landscape haunt dramas.

The fields of performed influences are naturally archaeological because both these fields of ghostly performance and archaeological fieldwork have similar characteristics:

- Both deal with material "bodies" (ghostly manifestations/ physical remains), material "things" (use of the sensory apparatus in telepathic communication/cultural artifacts), and physical settings (the haunted "mise-en-scene"/archaeological site);
- They both concern (and contain) experiences and memories; and
- They both make use of texts ("ghost scripts"/historical and ethnographic accounts).

These fields of influence in the anthracite region landscape represent, then, a series of potential archaeological sites <u>and</u> a haunt drama, both in terms of the social and ethnic actors who labored and performed in these fields of drama, and the fragments of those fields that still remain as sensory manifestations. The excavation of these landscape dramas is itself a performative act that requires a script to identify and unearth this drama. This field action requires a highly emotional and skilled performance on the part of the investigator that is both focused and controlled.

The landscape haunting dramas are very eidetic. They are visions of past actions that can be vivid and persistent (continuous). These may be recordings and/or entity interactions, and are at the "heart" of the landscape unfolding drama. They are what remains of lives "lived", and can take the form of individual (and highly personal) anomalous sensory manifestations. They consist of bits and pieces of what still is being acted out in a continued re-performance of a life, and what was experienced through the performance of everyday activities (habitual, mundane, and eventful). These performed actions influenced and determined the content of the haunt dramas that continue to occur.

These performances, and their excavation, are an archaeological attempt at explanation. A ghost excavation works within the symmetrical relationships between the past and the present. This archaeological fieldwork centers on:

- The question of agency and origin (who, script content, specific activity, emotional event, individual personality and author, and audience: the socially-significant person with whom communication is established);
- The question of remains (traces of residual or ghostly interaction; what types of traces; what performance is left); and
- The question of representation (presence or absence of anomalies; how should they be documented; how should they be represented in media presentations).

The ghost stories and haunting themes that follow are the anthracite region fields of drama that (may) remain as part of the contemporary landscape setting. These elements form part of the mise-en-scene of the anthracite hauntscape. This is a scripted narrative that emphasizes past-present relationships, one built on a coal-producing industry, rather than an exploration and documentation of what people said happened in history.

Ghost excavation is based on contemporary observation, recording and measuring, and making connections. The results of investigative and performed fieldwork are a different historiography. This is a ghostly ethnography of anthracite ethnic culture, composed of many haunting fragments of culture and tradition. This is meant to be a dynamic storytelling – not a narrative per se – between author (both past and present) and audience (also both past and present). Sometimes these roles are reversed, and the voices of the ghosts speak out, and we, the audience, must stop and listen. This is a different version of hauntings, and the sensory manifestations are not, for the most part, visible and easy to interpret. This is a re-mediation and re-contextualization of space because it is what really is out there, scattered throughout the landscape. Yes, it is woods and mines, abandoned coal structures, water courses, towns and patches. But it is more. It is so much more!

A ghost excavation de-emphasizes coordinates and temporal indexing because there is an innate symmetry to the land, and an unfolding of time that occurs here. It deals with "mess", that goes beyond the physical debris, because the categories of meaning are not tightly organized and, at times, appear not to be logical and practical. There is "noise", ruin, multiplicity, juxtaposition, ambiguity, multifunctionality, and non-function. It is both vertical and horizontal. It is a matrix of associations because there is continuity. The aim is simply to cause manifestation, as a controlled, performed, and contextual investigative exercise.

This is ghost excavation, and you – all of you – are welcome to participate and unearth the drama. Good luck and be careful. You never know what "dirt" you may uncover!

C. The Anthracite Region Landscape:
A Perceptual "Night Country"

Robert Lewis Stevenson once commented that some landscapes cry out for a story. The anthracite region is one such landscape. This landscape has been a continuous scene of perceptual discontinuity between a once virgin forest (a haven for migratory hunters), and the excavations of its later years. These excavations changed forever the landscape symmetry, giving birth to a large degree of haunting uncertainty. This loss of a Schuylkill "virginity" was a bloody transition, from life to death, and all colored in an ethnic red. This "death" held multiple possibilities, yet all were related to a mining industry that was responsible for the loss of that landscape innocence. The newly-arrived immigrant, working in the mining industry, "saw" death in numerous ways:

- You died working in these mines through the frequent accidents that occurred there;
- You died as a result of working in the mines through breathing-in the coal dust on a daily basis (anthracilicosis);
- You died because of affiliations to those who worked the mines (The Mollies and other labor-related murders);
- You "died" to forget you worked in the mines (excessive alcoholic consumption); and
- You slowly "died" because of the loss of work in these mines (economic decline and loss of jobs and income).

In this mining environment, numerous activities and events became marginal, including the humans who continue to live here. In this transitional state of discontinuity with the past, one became (or tries to become) something or someone else. Ties to the past became disentangled here, lost, and eventually forgotten. The once highly visible has become invisible because presence has become unimportant, and something that does not fit the current definition of "relevant".

In this new transitional state of slow economic decline, the physical remains continue to remain the remnants of a once-thriving coal industry (both human and structural), and the consequences of that decline. The most recent past became marginal, and the more remote past is obsolete. The past in general has become insignificant. It is what Loren Eiseley calls a "night country". The "night country" is defined and envelopes the mining experience: most of the time, the miner experienced his daily work routine, and associated activities, in the darkness of the mine. And this even affected the animals that labored there. Mules sometimes spent their entire lives in the darkness of the mine shafts, <u>never</u> seeing the light of day, or the warmth of the sun.

The darkness that was once an attribute of working in the mines has now surfaced and covered what still remains of that mining past. It is here in this "night country" on the surface where the ghosts of the anthracite region freely roam and manifest themselves. The investigative question is this: Are these interactive entities from the anthracite past, or are they merely haunting residual memories of a symmetrical past that is largely forgotten today?

One can enter this "night country" of the anthracite region, travel freely within its marginal zones and exit it, obtaining some interesting experiences in the process. The trick is to be open and receptive. That perceptivity is the orientation, the lighthouse, that helps illuminate the darkness that one encounters here, a "sea" of coal, soot, and culm. Without that illumination, what still remains will sink down and out of human sight, to submerge backwards into the liquid forest of yesterday, or the flooded (or fiery) abandoned mines of today. Is that why there are so many "sink holes" in this landscape? Is it due to our lack of insight into that past and its consequences?

This is the paradox of seeing "ghosts" in this "night country": to see, you need darkness. Too much light – too fast – and the past becomes merely a fleeting phantom. A proper balance needs to be maintained. One must understand the significance of the night in this landscape, and one needs to illuminate it correctly, aiming directly at the "heart" of these haunt dramas, as they unfold throughout the landscape and the mine shafts. After all, it is a spooky setting to work in. Anthracite mines "were spooky. That is why they were the source of so many ghost stories" (Korson 1960:306):

> "Where yonder yawning cavern opens wide,
> the car glides slowly down the rock's rough side,
> with trembling hearts we leave the upper light,
> and travel downwards to the realms of night,
> with wondering eyes we watch the sun-light die....
> Then, blind at first, we grope our onward way
> through paths ungladdened by the light of day....
> and changing lights and shadows mock our eyes:
> so strange the scene, awe-struck we seem to tread
> some fabled mansions of the silent dead,
> and start amid these shadowy haunts to hear
> the sounds of human toil salute the ear....
>   Thomas Llewelyn Thomas
>   "Coal Mines" (1863)

This human toil, together with the shadows that were perceived in the poorly-illuminated mines, helped to produce the ghosts that many miners believed haunted these mine shafts:

> "Many years ago the miners in the Greenwood Colliery near Tamaqua believed that ghosts of miners recently killed in a gas explosion roamed in headings and gangways".
>
> (Korson 1960:307)

A kind of cultural enlightenment is necessary to take us there and guide us through the night journeys in these mines and on the surface landscape. This cultural aspect nourishes the imagination by centering us squarely in the liminal space and time that lies between what is (yet past) and what could be (and re-occurs). This is the kind of guidance that leads us back through the same door we came in. And it avoids us getting "lost". It is the mirror image, reflecting our own "ghost within".

This is a view that penetrates through the veil (and vein) of everyday living by unearthing the marvels that lie in an excavated "night country". These "night country" excursions are <u>not</u> for everyone, or for every taste. The "night country" is for explorers who wish to transcend the harsh (and sometimes boring) daylight world of the reality that exists in this fractured landscape.

This night country was born in the mines, and came to fruition among the immigrants who worked deep in the ground. It flourished (and continues to do so) due to the symmetry between the Old and New Worlds. The mines and the night in both worlds were full of supernatural creatures, the least harmful of which were the ghosts.

160 years ago, there was no electricity, no illuminating guidance. When the sun went down, it became <u>really</u> dark. The only glow was a candle-lit luminescence, both in the home and in the mine. Shadows tend to accumulate and settle in this darkened environment. The silences, like sharpened knives, easily penetrated the night of mine and home. This darkness leaves a large blank space to fill-in. And it was. Both the miner (as a coal worker), and the miner (as a husband and father), filled in those empty spaces with stories in which ghosts and other beings from the imagination of their ethnic minds entered and thrived. These ghosts had a new permanent home, far from their European origins.

The stories continued to grow, unencumbered by the reality of knowledgeable daylight reality, and logical reasoning. A full day's work in the mines prevented this mental exercise. A house enclosed a family, and a mine enclosed a fellowship of co-workers. Inside these spaces, there was a rich vein of relationships and stories to help alleviate the daily stresses of life. When the door closed against the dark, or you are deep in the mine, the imagination opens, and the night country is awakened and begins to breathe life anew.

Yet, this night country is a vision of a landscape in daylight. It is night, not because of the time of day or the season of the year. It is a setting of darkness due to the presence of absence. It is there where these ghosts, largely unseen, can be heard more often than not:

> "There are songs and sounds in stillness
> in the quiet after dark
> sounds within sounds
> songs within songs

There are rhythms in the quiet
and pulses in the night
beats within beats
drums within drums

Something calling in the embers
something crying in the night
and out beyond the darkness
there are voices in the stars.

Felice Holman "Voices"

Which immigrants are speaking through the darkness? : is it the European, or the Native American? ; is it a coal miners pulse, or the drums of encamped native hunting parties? Whatever they are, they <u>are</u> the ghosts of the past that still continue to roam the landscape, and provide the sounds of the search for "game", and the search for coal. In the end, it is all one and the same because all of us, in this anthracite area landscape, are immigrants in the "night country" of the past.

Are you listening?

D. The Haunting Poetics of the Anthracite Region

"Where are even the words of that time, some lost, no longer spoken by living men"
                    Loren Eiseley, "The Mist on the Mountain"

The anthracite region is a "natural" and "sound" concept for a haunted landscape. It sounds haunted because the initial physical penetrations into this region were largely aural, not visual, engagements. In the impenetrable forest habitat, long before the "coal rush", there was a relative dominance of (and reliance on) the auditory sensory mode. This served as a distance-sensing devise (as well as a "game" locator) in terms of what was experienced and hunted. These sounds of the forest were subsequently organized into distinct mental images of animal life and other types of living beings. Here, in this forest habitat, there was a uniformity – and conformity – in sense and interpretation: an unknown sound was (could only be) "supernatural", be it the "spirit" of an animal, or some, still unknown, entity.

This tradition continued throughout the period of the later European immigrations, the "coal rush". The articulate aural symbolism of this spooky landscape, and its unique sounds, helped create a perception of "ghostliness". The auditory presence heard here also created ethnographies of ethnic sounds, serving as a cultural system of memory, remembrance, and the folklore of today:

- The sounds of the forest became the "wendigo" to the Delaware hunting parties;
- The sounds in the mines became "knockers" to the English miners;
- The sounds around the coal patches became the sounds of the "little people" to the Irish; and
- The sound of the wind through the trees became the "Baba Yagr" (the witch) to the Slavic immigrants.

In both the forest and the mines, vision was limited, and environmental sounds became the dominant mode of interpretation. These visually invisible presences formed the ethnic ghostlore of the region. It has been said that 60% of ghostly presence is auditory. If this is accurate, then the anthracite region is a very haunted landscape. The forest and the mines are full of "voices". Place, sound, and memories are symmetrically-linked here, creating a "surround sound" of haunting phenomena.

Here, in the anthracite region, there is also a deep relationship between the landscape and its cognition (image and perception) and the use of language ("coal speak"). This can be "seen" and heard in such words as:

- Braitch hole: This is not a tear in your new jeans. It is an abandoned coal mine tunnel;
- Breaker: This is not a person that destroys things. It is a building used to process newly-mined coal;
- Patch: This is not something used to "cover-up" something else. It is a group of homes attached to a breaker or colliery;

- <u>Stripping holes:</u> This is not a place to see nudist discard their clothes. These are abandoned mine excavations that have been used as "swimming pools" by locals;
- <u>Burma Road:</u> This is not an old road in Southeast Asia used during WWII. It remains a winding, but frequently traveled, road in the area;
- <u>Vulcan:</u> This is not the planet of Mr. Spock. It is still an inhabited patch in the area;
- <u>Culm banks:</u> These are not places where money is deposited. On the contrary, they are the "waste" of the search for coal;
- <u>Colliery:</u> This is not a person who breeds collies. It is an anthracite complex consisting of a mine and surface operation; and
- <u>Knuckle & rib:</u> These are not parts of human anatomy. The first is the curvature in the slope of a mine. The second is the support frame of an underground shaft.

These words (among hundreds more) are beautiful mixtures of landscape spaces, ethnic imaging, and language use that do <u>not</u> occur anywhere else in the world. So anthracite region resident, stop what you are doing for just a moment, put that beer down, turn that TV off a moment, take a pause in reciting that prayer, and reflect on that statement. And, more importantly, be more than a little proud. These are indeed auditory "treasures" of language use and sounds that no one else in the world can claim ownership to. This alone makes the anthracite region a truly haunting landscape!

"There was nothing exotic to name in that coal town, still we found words we liked the taste of and used them as we chose......"

Karen Blomain, "Mango"
(from <u>Coalseam: Poems from the Anthracite Region</u>)

E. Religion, Faith, and Apparitions:
A Phenomenological Perspective

"When the early settlers looked beyond the Blue Mountains and saw a great wilderness of forests, mountains, swamps, and streams, they unanimously decided that this was a realm of Satan and solemnly consigned Schuylkill County to the Devil"
William H. Newell, January 31, 1912
(read before the Historical Society of Schuylkill County)
(quoted from Korson 1960:432)

Schuylkill County, Pennsylvania is a region of tremendous religious fervor. Most of the small communities in the area contained a large number of churches of various religious denominations. Was that to offset the early perception of this region as part of the Devil's Domain?

A consequence of this fervor, and/or strong belief in its power and influence, has resulted in Schuylkill County, and adjoining areas, having its share of incidents of "religious apparitions". This can be seen in the sighting, by trapped miners Fellin and Throne at Sheppton in 1963, of Pope John XXIII in the mine, to the recent appearance (August 2007) of the Blessed Virgin Mary on a garage door on Lewis Street in Minersville. The occurrence of these religious apparitions has been attributed to the strong faith of the people in this region.

A belief in hauntings and the appearance of ghosts, as forms of cultural or ethnic (rather than religious) apparitions is a similar, though more mundane (and routine) phenomena. Ghostly presence, as opposed to a haunting presence, is the manifestation of a personality, in its cultural context, of a dead person, manifested before the living. This may involve one or many sensorial elements (smell, touch, sight, sound). If the ghostly presence was sensed visually, it was an apparition. An apparition is not <u>always</u> a "ghost". It could be another form of entity, a religious icon, or an image of an individual that was "created" by "natural" forces (such as high EMF fields, a recorded residual memory, or a hallucination).

In the latter case, it is called a "phantom", a vision or dream of the dead, rather than the appearance of the distinct personality of a dead individual. A "ghost" is still "grounded" to this world. A "spirit", however, has already "crossed-over" into the light, and has the ability to return, if only temporarily, to our world for a specific purpose.

The religious and the cultural apparition have a number of similar characteristics:

- A <u>shared</u> belief system (Christianity; "Ghost Culture");
- An observation of the world (and reality) through the use of the senses; and
- A strong faith in the authenticity of the visions that are being perceived.

All of these characteristics rest on a solid base foundation of the principle of phenomenology. Phenomenology is the study of direct perceptual experience through the use of all the senses. Reality is not a fixed point that can be analyzed and measured with scientific instruments. Reality is a symmetrical matrix of sensory perceptions. That is why a ghost excavation is useful <u>and</u> needed. The ghost excavation unearths the symmetrical layers of unfolding time and events in particular spaces, such that there is no clear separation (and distinction) between the past and the present, the living and the dead, a cultural or a religious apparition, a belief in something and its reality, and the cause and effect of each type of manifestation.

This symmetry connects these types of apparitions to one another, and all these types of "apparitions" are a form of amplified (and intensified) "ordinary" sensory phenomena. We perceive these apparitions when we "re-sense" the world through:

- Intensifying our relationships with "others" by understanding that the perceiver (us) and what we perceive are really the same; and by
- Amplifying the scope of our perceived landscape to include what we cannot readily "see" (and understand) because it is normally absent.

The history of the anthracite region has been characterized by a perceived absence. Yet this perception did not limit the scope of interpretation or explanation. The early Delaware hunting parties conceived of this region to be the forests of spirits. This area was located beyond the normal horizon of vision. It was a dense forest. It was called "Towanemensing". To the Delaware, it was an area where various animal species resided, and which were not always visibly evident in the observed landscape. It was a land where deer were believed to remove their animal guises and take semi-human form.

These were the "Manitou" of these early hunters, the spirits who inhabited these forests. It was part of <u>their</u> religious heritage. Was this the "changeling" I sensed in the woods, and described in <u>Ghost Excavator?</u> Did I hear its changing form take place: from animal to human, and back again to animal?

Later, different cultural and ethnic traditions (including <u>their</u> religious heritage) migrated to this area. But the end result was still the same: a perception of something beyond vision and verifiability, yet with human characteristics. This time, however, the mines replaced the woods, and became the new locations for these manifestations. The "beyond the horizon" concept was replaced by something beyond <u>and</u> under the ground.

The mines were the homes of ghosts and other beings, yet all of these entities retained their human qualities, and formed part of the social and religious worlds of these immigrant groups. These ghostly and spiritual realms were never fully cut off from the sensuous world of daily living. They were part of experiencing the world, however mysterious and hidden it may be. Its absence, most of the time, only made its appearance more unique and special, but not "supernatural"!

This intensity and amplification of perception still "lives on" today in the anthracite region in the form of both religious icons (Virgin Mary, local visits by future saints, the Pope and Mother Teresa) and cultural apparitions (ethnic ghosts). This means that we continue to have direct ties with the past, even though we are not always aware of these connections.

This perception of landscape continuity is what makes this area so unique. Why do so many people who live here not perceive this uniqueness? It <u>is</u> part of our rich cultural, ethnic, and religious heritage. Remember this the next time some "outsider" calls this area a "depressing" place to visit! The religious and cultural symbolism of this area is manifesting everywhere. Rather than depressing, it should be a source of joy for both the living <u>and</u> the dead!

F. The Photographing the Dead….Again: Memories that Last (Beyond) a Lifetime

Schuylkill County and its environs is a <u>great</u> place for photographing the once and future dead, and you don't have to enter any cemetery to see any ghosts! They are clearly seen walking the streets in the local communities today. I am not suggesting a place of ruins and walking zombie-like beings. Rather, these are places that are <u>still</u> alive (however abbreviated) with the rich and varied traditions of the past. This is ethnic photography. But it is also more – much more – than photographing events (weddings, social/family gatherings, and reunions) or eventful activities (church and fire company bazaars; community days).

It is the preservation of a moment of the past that belongs to a specific social situation, even if the moment is a landscape portrait. All photography is conceived and executed in a social environment, and under circumstances that are <u>now</u> completed, <u>but</u> still continue because they are remembered and re-shared, if only among family members. These photographs become a symmetrical part of the contemporary environment. This <u>is</u> a haunting experience, and its "ghost" (the remembrance) remains "alive", though not physically still breathing.

This situation speaks more than particular "captured moments", such as the intentional search for "orbs", photographing them and other light anomalies, characteristic of many "ghost hunting" field strategies.

An orb is not a photograph of a social situation. It is not illustrative of context, and thus is not significant. What event or activity does an "orb" photo really "capture"? Is it a dusty or humid environment (which is true in most cases), or something else? This photo may have been a "lucky" shot, or framed by a "reading" anomaly. This "reading" (or a measurement of an ambient environmental anomaly) is not social, and it may not even be communicative, in any "sense" of the word.

Cultural and ethnic photographs are a means to enter the Shakespearian "undiscov'd country". To photograph the "dead" is to bring a re-birth (to re-animate) something that <u>is</u> present, but is absent from our contemporary visual memories and thoughts. It resurrects what has become a "ghost" of some former significance, if only for a fleeting moment.

We can make this ghost currently active by photographing it. A photograph remembers its origin, and without bias. A good example of photographing a "ghost" is the St. Nicholas Breaker, located between Mahanoy City and Shenandoah. This structure has been photographed many times since its "death" as a functioning coal breaker. The vast majority of these photographs, though, have been made by contemporary bands of "hunters" who temporarily migrate here to "shoot" their prey, the breaker. Doing this, they mimic the Delaware hunters of old, and, like those Native American hunters, after "capturing" their prey in sufficient quantities, they return to their own lands to partake of the fruits of that hunt. In their hunting expeditions, they take <u>our</u> memories with them. We are losing our "dead" to these others, who do not understand (or care about) the society that gave birth to those social memories that are represented in these now-abandoned structures. Our forefathers lived their daylight hours in such structures as the St. Nicholas Breaker. <u>We</u> are connected to those lives; not so, these non-local photographers. This photographic exercise is part of <u>our</u> ethnic tradition and history.

The St. Nicholas Breaker

Another Image of the St. Nicholas Breaker

The St. Nicholas Breaker in its "heyday"

At one time, it was an Eastern European tradition to <u>really</u> photograph the dead. The symmetry of past and present, in these photographs, became a future memory. This particular form of representation of the "ghosts" of the past is a personal memory for me. I once saw, and held, those types of "ghost photos". They were taken at the funeral of my Uncle Stephen, who died very young so long ago. Around his casket were my grandmother, "Baba", my Aunt "Ceci", and my dad (an infant) among others. That photo is now "lost", somewhere in the storage of time, a ghost of a memory that still haunts me even today.

Let us not lose these "ghost photographs" of the past. Let us reclaim our ghosts that still haunt and inhabit this landscape, our landscape. Let <u>us</u> photograph these ghosts, and keep them haunting <u>our</u> memories. Let us pose together (a tribute to that Eastern European tradition of long ago) in front of these towering "dead" entities, if only to remember (for a time) that captured moment that was part of a (our) shared social situation and history. Do not allow these phantoms of the landscape to migrate away from the homes (and a land) where they <u>really</u> belong. That migration ended long ago with <u>our</u> grandparents and great-grandparents. We, and our memories, are now part of this land. And that is the contemporary "ghost hunt" of this anthracite coal region!

The Excavation

A. The Area Hauntscape

A.1 The "Call of the Wild": The Anthracite Aural Presence Today

"Have a care, even in the symbol-shifting brain we may be unable to escape prophetic things. We may be wandering on our way on the roads of night, hearing the howl, the guttural laugh of that which will replace us"

Loren Eiseley, "Why Did They Go"

"No sight in the forest is without significance, not a glade, not a thicket that does not provide analogies to the labyrinth of human thoughts. Who among those people with a cultivated spirit, or whose heart has been wounded, can walk in a forest without the forest speaking to him"

Honore de Balzac

To drive the back roads at night in the anthracite coal region, admiring the "silences", the blackness of the culm banks and alternating woodsy and barren slopes – mostly untouched by contemporary mind excavations – seeing the stillness there is, is to re-envision, perhaps, the metaphysical fauna of Delaware hunting band lore. It is sensing the little forest people of the Irish immigrant, or the bipedal sacred deer, the Manitou.

Spirits seem to filter through this land, escaping into secret passages other than the now-abandoned mineshafts. To actually walk this landscape, along the lonely paths, is to breathe in their air, and feel them anew. This is the symmetry the landscape contains:

> "By midnight moons, o'er moistening dews,
> inhabit for the chase arrayed,
> the hunter still the deer pursues
> the hunter and the deer, a shade!
>
> Philip Freneau (1752-1832)

These are the ghosts that still roam the back roads and woods at night. Yet, when one tries to make sense of these fleeting encounters, and where evidence is nearly unattainable, do not expect more. This is not an easy journey, or an excavation without heavy labor. Be happy and content with what you may have already glimpsed, for these are the haunting uncertainties of this land, and only occasionally surface through the accumulated layers of memory. Is this partial vision enough to explain the ghostliness? That is the problem when one deals with a spooky landscape. We are left in doubt.

This anthracite region is still a "wilderness". This wilderness is not dependent upon a vast, unsettled track of land. To Thoreau, it is a quality of awareness – rich engagements and openness to the senses. In this essence (or lack of it), the anthracite region landscape remains the Towanemensing of the Delaware hunters, because it is still largely unexplored by those who live here:

"I am the land that listens I am the land that broods steeped in eternal beauty, crystalline waters and woods long have I waited lonely, shunned as a thing accurst, monstrous, moody, pathetic the last of the lands and the first...."

Robert W. Service

There is an alternative way of looking at something that is not so easily seen. It is in the form of a rediscovery of what we already have, what already exists, but somehow, because the voices were not heard, has eluded our recognition and appreciation:

"Yes, they're wanting me, they're haunting me,
the awful lonely places,
They're whining and they're whimpering
as if each had a soul...."

Robert W. Service
"The Lure of Little Voices"

Instead of being a history of what was, this is an exploration of what may yet be encountered hidden beneath the layers of the mundane, the habitual day to day activities that can be, when "tuned-in" properly, become the auditory residuals of a past. This is an excavation, meant to recover the veins of myth and memory that still lie voiced in the mineshafts, and heard in our minds. Here, it these largely unheard caverns, there is a rich tradition of ethnic and Native American folklore.

To see the haunting presence beneath the ghostly outlines of abandoned structures is to be made aware of the lasting spookiness of the core landscape. It is centered in a place which periodically exposes its ancient visions of forest, and underlying caverns of imaginative things. An excavation will uncover bits and pieces of an apparent mythological design that can lead us deeper into the past, as we dig down through layers of memories and representations of enacted experiences exposing, at the surface, the symmetry that is past/present space.

We accomplish this through hearing these voices. To not to listen to these voices, is a question of an under-exploitation of this anthracite landscape. To not do so will forever delegate this land to remain in Shakespeare's "undiscover'd country".

The mines and abandoned structures, including uninhabited houses, are appropriate fields for hearing these voices. The surrounding forested areas, visited by the occasional haunts of today's hunters, are another field of ghostly drama. These areas still remain "underground", unexplored "patches" that dot the countryside of the anthracite region. Even natural sounds coming from these "underground" areas affect both humans and their culture. In fact, it has been suggested that noises generated by subterranean ground movement, water, wind, and wildlife were a reason why Mesoamerican cultures (Aztec, Maya, Toltec, etc.) perceived caves, limestone sink holes, and forested areas as sacred (Bruchez 2007). Though anthracite miners did not conceive of the mines as a "sacred" place, they were considered the abode of supernatural forces, as was the once great tracks of forested areas (by the Native Americans).

We can gain access to this wilderness of "underground" areas here by becoming more than a "tourist". Its exploration requires a "Faulknerian" view of history, and its participants: "the past is not dead, its not even past". An immersion into this "underground landscape" is the embodiment of the probability of encountering ghostly images, recollections of memory and experience, and the "little voices" that call to us in the forest. These are very emotional haunting uncertainties. We are on a time machine back....to a past, present still. In the anthracite coal region, the dead are not really dead. They are waiting, still located in that vast "underground" wilderness:

"There's a land where the mountains are nameless;
and the rivers all run God knows where;
there are lives that are erring and aimless,
and deaths that just hang by a hair.
There are hardships that nobody reckons;
there are valleys unpeopled and still,
there's a land – oh, it beckons and beckons,
and I want to go back – and I will"

                                        Robert W. Service

The land was (and still is) the land of Towanemensing, the "underground wilderness" of the anthracite coal region.

A. 2. The Forest of the Peddler: A Grave Concern

H. J. Massingham, a writer of matters of the countryside (among other interests) once remarked that the woods of today (early 20th c.) are not haunted by ghosts, but by what Loren Eiseley has called "a silence and man-made desolation that might well take terrifying material forms" (1969:26). I, myself, personally experienced those material forms, which were described in <u>Ghost Excavator</u> as the "changeling in the woods". Was that a "shape shifter", embodied in me, or some spirit entity in those woods?

Changelings may be a product of the "perceived" immensity of the forest. This may not have anything to do with physical geography, at least not for these anthracite region woods. It is in a cultural landscape that this perception evolves, and these local forests are full of cultural manifesting elements. It is the sense of one going deeper and deeper into these woods, eventually becoming "lost" culturally (and mythically), and being no longer tied to 21st c. roots. This cross-cultural immersion connects one directly with the past, by eliminating the barrier between a visual present, and a past that still continues to exist in these anthracite woods.

We must immerse ourselves in the sounds of the past, as they echo through the trees. This sense can be heard clearly, if we allow the visual to fade in importance, and the auditory to take precedence. It is in this state of auditory awareness that the terrifying material forms, mentioned by Eiseley, can take physical form, and the sounds of the peddler, Jacob Folhaber (who was murdered in these anthracite woods in 1797), can be heard, without even seeing his presence. This happens because this is his forest, still:

"Pious forest, shattered forest, where the dead are left lying
Infinitely closed....on the vast, deep, mossy bed, a velvet cry".
Pierre-Jean Jouve, <u>Lyrique</u>

The woods near the Peddler's Grave

The ghostly presence of the peddler is a cultural haunting. Cultural hauntings are the result of a sense of loss of the history of cultural and social life in a particular community:

"Ghosts operate as a particular and peculiar kind of social memory, an alternate form of history-making in which things usually forgotten, discarded, or repressed become foregrounded...."
(Richardson 2003:3)

Unfortunately, today the peddler's cry is seldom heard, or even remembered. His presence is no longer felt by most people. Instead, these woods of the peddler contain "no trespassing" signs. These symbols loudly speak for the state and loss of anthracite cultural heritage, by continuing a history of the few, who monopolize the memory of the struggles of individual presence. The signs are clear: do not come here, you are not allowed to enter these woods, our woods.

These peddler woods are "sacred", not because of their naturalness or the presence of "game", but by virtue of their memories. The murder of the peddler was significant. It was the first known act of violence (and subsequent justice) in an area that became frequently violent during the era of the "coal rush" (1840-1910). For that reason alone, these woods should be treated with symbolic memory, and the remembrance of a violent past.

We are exposed to that history (and its violence) by becoming "sensitive inhabitants of the forests of ourselves" (Jules Supervielle). The death of the peddler, forgotten by many, only resurrects itself through a ghost excavation, and not the desecration of his grave marker with beer cans. But knowing the history is still not enough. We have to experience it, both through others and ourselves. By recalling the past memories in the same physical space, we can expand backward both the history and a physical connection to the original event. This occurs even though the physical features may have been altered (these woods are not the same as they were in 1797). And by re-performing the event in the same physical space, we may unfold that past back into the present (see below).

The woods become a heritage forest containing a sense of place that may extend backward through several lifetimes. It is a forest full of material forms and hauntings from the past whose human temporal depths may reach back several hundred years, if only in fleeting glimpses. All of these multiple pasts are characterized by their migratory nature. In this forest of symbols, man, both Native American and European, is the transient. Through their migratory incursions, this virgin forest slowly became an excavated space, by the hands and minds of man. Unfortunately, much of that past was (is) violent (from hunting "game" to hunting humans to hunting "game"). That alone preserves the memory of these immigrant woods as a place of haunting uncertainties.

A. 3. "Play Station" at the Peddler's Grave

"Whenever the moon and stars are set,
    whenever the wind is high,
all night long in the dark and wet,
    a man goes riding by.
Late in the night when the fires are out,
    why does he gallop and gallop about"?

                              Robert Louis Stevenson

Is this man, Jacob Folhaber, the headless horseman of the anthracite woods? Does he still ride his ghostly horse through those darkened woods on the anniversary of his death? Are those rustling sounds in the surrounding brush merely the "natural" sounds of some nocturnal animals (a horse, perhaps?)? Do we depend too much on "wishful thinking", the power of imagination, or the suggestibility of this early anthracite ghost story to create a haunting image?

Whatever (or whomever) is riding in those woods (or in our imagination) is better, and much more detailed, than any simulated computer game that is played in the comfort and safety of home! Besides, isn't imagination a far better mind "tool" and exercise devise than sitting in front of a video screen and merely playing "make believe"? Isn't the video game a duplication of a haunting: something that is "played" over and over again? Go to the woods around the Peddler's Grave, and play the game of "hide and seek" for real! It's a "live" ghost hunt!

The site of the Peddler's Grave "ghost hunt"

But like home video games, there are rules to follow. Here are some rules for our "ghost hunt":

- Seek permission – do not arbitrarily trespass;
- Do not go alone (for safety and security), and take a cell phone;
- Do not litter or trash the place, especially with beer cans. (remember no smoking or drinking during an investigation);
- Respect the gravesite, and the person it represents (Remember: he probably is not buried there. But his ghost may still wander the woods. At the very least, his residual energy is still there, a recording of this highly emotional event);

- Visit the place first in daylight – know your surroundings and the hazards it may contain;
- Perceive the natural details of the area. At night, these may not appear the same;
- Be open-minded and conscious of, and sensitive to, the sounds of the woods – especially the natural sounds;
- Be active – not passive. Do not wait for something to happen – do things! Be contextual in your approach. Be a little knowledgeable. Research the history and details of the murder. Talk to Jacob, using some German phrases;
- Be aware of the sensuality a haunt drama can manifest. Think what he was doing <u>before</u> he was murdered? What was he thinking? ;
- Be patient! Do not approach the site as a "once and done" type of investigation; and finally,
- Identify with the German peddler as a "person", not an "object" for fun, fright, or financial gain.

The most important factor in this "gaming" is to be socially-identified as someone a "peddler" (and not a "ghost") would communicate with. This image of a person, not an investigator (or a "thrill-seeker"), will open the lines of sensual dialogue with the past of these anthracite woods. Be a player, the right type of player. GOOD LUCK!!!

A. 4. St. Joseph's: The Cemetery of Dead Lithuanian Memory

The St. Joseph's cemetery is a fragmented, and abandoned space of burial plots and their toppled (in some instances) memorials. These grave sites are, in the unattended confides of vegetative growth, conjoined together, perhaps forever, by a frail network of neglect and forgotten heritage. The dead are linked, as in life, by their ethnicity.

The abandoned burial plots of the St. Joseph Cemetery

More Abandoned Plots at St. Joseph's

These dead Lithuanians are "lost" again, left to explore a new and distant world, this time, in stoic silence. The compass that orients their passage, covered in weeds and thick brush, obscures their journey. These are the ghosts of cultural hauntings. When will these inhabitants of the grave, and their past dramas, become symmetrically intertwined and re-immersed into the world of contemporary Lithuanians? Both the fore bearer and the descendent face barriers of extraction from a past that has somehow lost its relevance today, and, more importantly, the lessons that have already been learned have been abandoned or ignored in the weeds of eternal growth.

These neglected graves belong to different times that involved stronger social ties and relations. Yet, their remains are physically present today, although in a different world than they remember. Or is it really so different? This possibility connects and unfolds the two time frames (the living and the dead) to a mind set that is based on loneliness: a place (this cemetery and its anthracite landscape) that has fallen out of proper service and function to the world at large, and in a society in which both continue to inhabit.

The "residents" of St. Joseph's are "true" ghosts, faded and forgotten fragments of history. They remain, culturally haunting this landscape. Unfortunately, so too are those contemporary ancestors who fail to remember their presence here. The absence of care or concern for history, heritage, and family is the real haunting that is experienced here.

At this "old" St. Joseph's Cemetery, the sad reality is that one does not need to excavate this particular landscape to unearth the "dirt" here! It is seen daily by all those who drive past the entrance, and do not remember, or recall the lives, of those who still lay buried there, forgotten in time.

There are stories of ghostly presence here, and there is a growing folklore of contemporary myths about this place. One of these ghost stories is that, on certain nights of the year (conveniently unspecified, thus not enabling one to properly investigate and document any manifestations), the caretaker's shack, located to the rear of the burial plots (and away from the road), and the associated stone wall, "disappear". Is this symbolic of the fact that the caretaker is no longer needed; or a haunting memory of what this cemetery lacks in the 21st c., such as a proper landscaping? Or, is it a reference to what occurred in 1922 with the burial of a young girl, mentioned in Ghost Excavator?

The St. Joseph's Caretaker Shack and associated wall

Energy is scattered throughout the area here and, at night, digital photography <u>will</u> reveal light anomalies or "orbs". Are these really manifestations of Lithuanian ghosts, or the result of high water moisture in the cemetery? Go out there and take some photographs yourself; but use a 35mm, instead of a digital camera. See what develops. If you go, and before you leave, say a little something to (and for) those interred there. Pour a little "boilo" on the graves. They deserve it! It's best to do this on November 2. It is "velines", the ancient Lithuanian holiday, the feast of the dead.

In the cemetery, especially in the summer, the area is thick with vegetation. Trees are everywhere, competing for living space, and thus "robbing" the space of the dead. Yet, this is appropriate. The ancient Lithuanians believed that the souls of their ancestors, which would include the dead buried here, "lived" in the trees. Is that why the burial plots are "dead", and the area is "alive" with trees? Together, the burial plot and the tree are conjoined in a society in which both the living and the dead continue to live, and inhabit…..

A. 5. The Silent Coal Industry Structures: Are They Voiceless or Symbolic of Something Else?

"There is nothing like silence to suggest a sense of unlimited space. Sounds lend color to space, and confer a sort of sound body upon it. But absence of sound leaves it quite pure and, in the silence, we are seized with the sensation of something vast and deep and boundless"

Henri Bosco, <u>Malicroix</u>)

Abandoned coal structures near the St. Nicholas Breaker

Another Abandoned Coal Structure near the St. Nicholas Breaker

The silence of these (now) unused structures creates a haunting atmosphere. These structures, and especially the breakers, are the most characteristic feature of the anthracite coal region, and today, is its most prominent and visible ghostly presence. The "breaker" had its beginnings at the Wolf Creek Colliery, north of Minersville, Schuylkill County, Pennsylvania (Korson 1960:90). This occurred in April, 1844. The coal breaker itself was invented by Joseph Battin in Philadelphia (Ibid: 90). These breakers were used to "break-up" the coal into standardized sizes for shipment. The rest of the debris was dumped onto dirt banks, called "culm banks". These "culm banks" were also the locations of ghostly presence in this region (see below). Sometimes, the associated culm banks loomed higher than the breaker itself. What a waste! : discarded material more accumulative than useful. Today, these culm banks are becoming useful by providing "fuel" to the "co-gen" (coal generating) plants that dot the landscape.

How do we unearth this haunting atmosphere, and separate it from the ghostly entities and their individual dramas? What is required is an excavation to determine this ghostly presence. Dare we unearth the truth – uncover the "dirt" – why these structures were <u>really</u> abandoned?

What still inhabits these spaces of silence? Have <u>you</u> heard the echoes resonating throughout these structures, like the ball in an old arcade game? Was that someone calling from the past, or merely a passing bird in flight (or fright)? Were those footsteps that were heard ascending (descending) the stairs? Yet, the continuous sounds of rhythmic approaches do not betray the fact that there are three missing steps! Stairways are <u>very</u> haunted spaces! Did that shadow, seen out of the corner of your eye, make a sound; or was that the sound you made, as you perceived that shadow?

Here, in these abandoned structures, there are many possibilities, each one sounding sane or insane at the same time! Do we have probable answers, beyond the imaginative? Do we have a "ghost of a chance" to learn the unwritten history and drama of these buildings, and exorcise the haunting uncertainties? Should we even try?

The answer lies in whether we want these structures to become part of a rich anthracite coal heritage, or merely a "side-bar" to history, a ghostly memory, largely forgotten, of the past. YOU DECIDE!!

A. 6. The Culm Banks: The Creation of Future Ghosts – And A Potential "Hell" on the Surface

A Culm Bank

The fire in and around Centralia is a well-known and documented subterranean phenomena that has finally surfaced, revealing its true presence. But there are other fire hazards in the immediate surface area of the anthracite region that are equally threatening.

Culm banks are large deposits of waste materials from the mines that are scattered throughout the landscape. Sometimes, they catch fire, either by spontaneous combustion or even lightning strikes. In

1970, for example, there were 30 anthracite culm banks (and mines) burning in the state of Pennsylvania (Dinteman 1995:5). Though these culm banks are diminishing, many still remain. Who will put out their fires if they ignite? What volunteer fire companies will respond? All of these ghosts of a once-thriving industry are located outside of the towns and patches. Who will respond to this potential hell on earth? Certainly, not the "ghosts" which created them! :

> "Mountainous waste piles of slate and rock, black banks of coal refuse that stand stark and forbiddingly barren, permitting a scattering of low shrub and scrawny trees with a few black dust covered leaves. Mountainous black banks of coal refuse some turned copper-color from fires within that burn on and on and on".
>
> Walter L. Dinteman, <u>Anthracite Ghosts</u>

Do we need a "ghost hunter" to exorcise these potential phantoms? Or is the danger only a haunting memory that has past? Only the ghosts know with any certainty what (and where) something, and that "fire", <u>will</u> happen next!

A. 7. The "Coal Dust" Ghosts

"This quiet dust was gentlemen and ladies, and lads and girls
– was laughter and ability and sighing...."
> Emily Dickinson, <u>A Cemetery</u>

Here, in the anthracite region, we carry within our bodies all the minute fragments of the history of this landscape. This is the coal dust particles we breathe daily, and that circulate freely (as opposed to the miners who worked the mines) through the air here. The culm bank deposits, and the dust "orbs" photographed in the abandoned coal industry support structures, are the visible outward remnants of that past. This history is also contained in the graves of our dead, as they pass through the once worked mine shafts, and are the hidden (and ghostly) memories within ourselves. These "dead-cold" ghosts are always with us, but they may go unnoticed, and become lost to our contemporary sensitivity, in the reshuffling of "outsider" feet, and into <u>their</u> lungs.

There is a long history of these coal dust ghosts, and that history began early in the individual lives of many breaker boys, who became future "miner" ghosts. These breaker boys had a "baptism of fire" that began the process which led, if they had an unfortunate accident, to their ghostly presence in these mining structures, still dotting the landscape. This "sacramental" rite consisted of kicking a cloud of coal dust down onto a newcomer's head from the rafter's above – a dust to dust ritual that became part of the "last rites" for these breaker boys. (for more stories of the breaker boys, see below)

The once virgin forests of the region were indeed "virgin", initially devoid of these coal dust ghosts. The birth of the "coal rush" gave rise to the birth of the coal dust ghost. In the mines, the circulating (and stagnant) air was the maker, (not by any divine deity or supernatural being), of these ghosts. If you worked in the mines, you breathed-in that coal dust – and it filled your lungs. That daily dosage of toxic material led to many premature deaths, and other health complications. Thus began the haunting dramas, and the origin of these coal dust ghosts.

This same air is still circulating, still haunting us, in these abandoned surface mining structures, operations that once "breathed" both life and death into the local economy. But now, the ghosts of the past are the only visible (and audible) voices of that past that continue to inhabit these breakers, collieries, and mines. They are visualized, to the amateur, as ghost "orbs" recorded on digital cameras by the "ghost hunters". It shows how the perception of the past is merely a grain of dust particles that still influence the present, irregardless of the misconception these "orb" photos may produce to a willing and gullible audience.

Sometimes, however, the thickness of the dust particles produces fog-like apparitions that seem to come out of nowhere, or is it rather somewhere in particular? Is there still activity in these abandoned structures, and in these deserted mine shafts? Do dead miners, lost in "cave-ins" and other accidents deep within the mines, still continue to work their shift? Have real ghosts replaced these coal dust phantoms?

Does the coal dust itself from the mines and structures provide an avenue of transport, and a channel of communication, for these dead memories of working conditions from the past, allowing them to circulate in contemporary space?

The answers to these questions is a personal vision quest, to those inclined to explore it, one that is buried deep within the minds of each one of us who live in this mined-out landscape. The quest requires an unearthing of these mined and excavated spaces. The haunting reality is (and this is the ghostly certainty) we are all united together as one in this landscape. Our bonding agent can be found in the coal dust. Understanding this is the beginning of <u>our</u> "rite of passage" into this haunted landscape.

A. 8. Nights in Haunted and Abandoned Anthracite Buildings

How many locations in the anthracite region are haunted? The answer (potentially) is: as many as one has spent at least a "night" (and day) in! All sites have, at least, some residual recordings. How many of these residuals contain an interactive presence is a work in progress by the C.A.S.P.E.R. Research Center. However, this investigative work may take more than a <u>single</u> lifetime to complete, since we have to include "dead time" in the equation, and multiple occupations of symmetrical space that will continue to "accept" deposits of haunting remains in the future.

In these haunted (and occupied) places, the surface debris of a continuing drama still grows, accumulating memories, and then unfolds. There is a sense of unmasking at these locations, such that the underlying dramas do not readily surface, thus hindering the exposure of this "something else". The unfolding of the past waits patiently, since time is not a consideration, before it can re-occur (and re-appear) in the present. Ghost excavation reveals the process of unearthing, and that is what nights at perceived haunted locations are all about: to seek out those who chose to remain, or could not enter the "light" (after too many years living and working "in the dark"). How many former residents remain in these structures is one research question (and one of the investigative goals) "asked" by C.A.S.P.E.R, during our nights in the "haunting zones" of the anthracite region.

These locations are the places of transition where light and noise can be re-exposed, after the reality of the darkness and silence in the mines, had ended. The miner was once a form of "living vampire" who was accustomed to be "active" in the dark, especially during the long winter months. Death, to these miners, was a true "paradise", as he could then enter "perpetual light" (if he so chose). That is why the miner cemeteries were located in the hills, away from the town and mines, the places of daily darkness and struggle. Here, on the rolling hills overlooking the towns, they could rest in peace, and in the full light of day. The places below, in those mines, was the hell that was lived daily under the earth. Were they already the "living dead", buried in those mine shafts, yet working day after day?

At death, there was a time to celebrate with drink. Is that why many funeral "meals" were (are still) celebrated in the restaurant/bars of the region? Is that why there were so many bars in the towns of the anthracite region? In these many bars, they serve more than "spirits". They still pour drinks for the "miner" ghost! During the heyday of the "coal rush", did the miners celebrate (after a day in hell) another night closer to death and everlasting light?

If so, who then are the ghosts that still inhabit these coal towns? Certainly, they shouldn't be the miners? Those that still remain, at least in the haunted houses in the towns, are most likely those individuals who did not toil daily in these mines. And that is why, giving the hardships endured by the miner, these non-miners that still remain, spend most of their days (and nights) toiling in these homes, rather than in the mines. The "nights" and darkness of their ghostly presence, and their routine and mundane activities, become also the days of their manifestations. The repetitive nature of their home-bound activities has become the routines of their haunting dramas. And it is this routine pattern that the investigators of the C.A.S.P.E.R. Research Center encounter in those nights in haunted and abandoned buildings in the anthracite region.

A. 9. "Breaker-Style": The Young Gothic Ghosts

The breakers and collieries of the anthracite coal region are a distinct architectural (and visual) design. They are the anthracite version of those medieval Gothic cathedrals that are scattered throughout the European landscape. These coal industry structures, and their European counterpart, the Gothic cathedral, share the same sensual representation. They both symbolize a particular landscape setting, physically and culturally: toil and trouble, mixed with deep faith, and a hope for a brighter future.

All that remains of these anthracite "gothic" structures are gigantic and decaying hulks of presence. Like the Gothic cathedral before them, they were constructed from large amounts of labor and toil. These symbolic structures created the "religious fervor" called the "coal rush". For a large majority of the mine workers, it was not the immigration to a new life, but rather a "rush" to a quick, and painful, death. This haunting reality created the anthracite ghost, and are a product of their culture – their ethnicity and values – creating what is called "cultural hauntings". These cultural hauntings are deeply meshed in religious overtones, first Protestant and later Catholic. They are the American counterpart of the Gothic-inspired religious fervor of Europe during the Middle Ages.

The Gothic designation seems an appropriate concept for these "coal rush" structures. Once majestic, and fully-functional and working facilities, they are characterized by their vertical physicality (most are several stories high), and their expansive areas of penetrating light and solemn shadows.

At the height of the "coal rush" in 1917, there were over 300 breakers in the anthracite coal region, and were the visible symbol (like the Gothic cathedral) of economic and religious (the daily "rituals") activity.

These breakers were structures, located in liminal zones between life and death, because they took the life blood of those who worked there. This blood was transformed (for more than a few) into the death agonies of potential future ghosts. In the November 1906 issue of <u>Cosmopolitan Magazine,</u> the coal breakers were described as "high-piled dumps of culm....(that) stretch in lifeless table-lands around....that are forever vomiting forth more dead matter to stifle the discouraged life of all green and growing things".

Sometimes, this dead matter included parts (or even whole bodies) of a breaker boy! In these breakers, much of the workforce was composed of young boys, the "breaker boys". Surely, if any ghosts still roam these structures, it is these "breaker boys", still searching for their manhood....and childhood dreams! As the coal came down the coal chutes, the breaker boys sorted through the coal, picking out slate and rock. In their daily work, the breaker boy relied heavily on their sense of touch. They were not allowed to wear gloves. The coal debris, rubbing constantly against these fingers, made them bleed. These bloody fingers, even today, can still be felt in some of these older breakers, if only in the memories of their sufferings. Sometimes, a lack of attention and immaturity, at their tender age (some were only 8 or 9 years old), often put them at great risk. This, coupled with their "spirited" nature, sometimes led to their untimely death, crushed by the machinery that broke the coal into smaller pieces.

In another time and era, the small boys that assisted at mass in the Gothic cathedrals were clean and clothed in white, and very different from the boys in the latter-day "cathedrals" who helped separate the slate and rock from the coal in the anthracite breakers. The "altar boy" and the "breaker boy" were two distinct individuals, separated by more than geography and history. The anthracite breakers not only broke the coal, it also broke these breaker boys in their daily work. Instead of white cossacks during a mass at a Gothic cathedral, only the ever-present black soot, mixed with ethnic blood red, accompanied these breaker boys on their daily "chores".

The breaker boys learned four things in an anthracite coal breaker: how to chew tobacco, smoke cigarettes, how to acquire a rich "curse word" vocabulary, and how to haunt the breaker where they worked. If they survived the ordeal of work, they had already amassed a culm bank of memories of the horrors of the work experience. Sometimes these memories would last beyond the grave. If ghosts do indeed exist at these coal breakers, they are the ghosts of the breaker boys. The "ghosts within" these boys (coal dust, ear-piercing noise, backaches, bloody fingertips, and the chew tobacco and cigarettes) may have prevented many of them a restful, eternal peace, and forever chained their "spirits" to these breakers.

If they survived the breaker experience, many of these same boys became future miners. In that evolution of work responsibility and labor, they were returning to the "dust", even before they were "officially" declared dead! They were literally digging their own graves. The dust in the breaker became the dust in the mine. Both "dustings" led to sickness, and, for many, an early death. Was that the "bright future" these breaker boys hoped for?

Working in the mines often led to a quick trip to "paradise", physical death. Death in the mines was so commonplace that the "Black Marias" (horse-drawn ambulances) was a frequent experience (both visually and aurally) of the anthracite coal region. Interestingly, white ladies and the black marias are frequent "feminine" haunting presences in this male-dominated work environment. At dusk, on moon-lit evenings, do the black marias and white ladies continue to manifest near these coal structures and the old mines, waiting to haul away the human debris (again) of a coal experience tragedy? Take a walk out to these old remnants of "King Coal", and go beyond the safety and illumination of the coal towns and patches, and discover for yourself this macabre aspect of the coal region's bloody past!

And that is why these breaker boys, especially on cold wintry nights, can still be heard in these breakers, as they slowly (and painfully) separate slate from coal, and renewed memories of life from their deathly existence. The breaker boy, in this "American Gothic" setting, is the "gargoyle-shaped" ghost that haunts these anthracite cathedrals.

Breaker Boys

B. The St. Nicholas Coal Breaker

B. 1. The St. Nicholas Breaker: A "Phantom" Structure That <u>Still Remains</u> "Alive" in the 21st c.

The St. Nicholas Coal Breaker, and its surrounding hauntscape, is a "loaded" setting, a place of historical and ethnic presence, yet one containing a cultural absence. This is a direct result of visitor penetration, those who come here to view this structure for personal gain. In most cases, it is an "outsider" view, not a local gaze. This visitor culture is not part of the Anthracite ethnic culture that spawned the breaker. This is not a bad thing in itself. It does create interest, albeit monetary. What is critical is that there is a monopolization of these "outsider" visits, rather than local interest or concern. As I pass by this structure, almost on a daily basis, the deterioration of its frame, and <u>whom</u> it represents, is rapidly de-composing into oblivion.

Yet, there remains a series of connections in time and space. Even though the visitor is a temporary "immigrant", he (or she) continues to follow an historical precedent like many others who came before, such as the Delaware hunting bands that traversed the region (however fleeting) in search of game. These "new" temporary visitors also seek "game". These are the photographs they take of the breaker and its environs. These visitors are the real <u>contemporary</u> "ghosts" of the St. Nicholas Coal Breaker. Today, they are seen as fleeting glimpses, a theatrical, perhaps even comic, presence, as they scurry in and around the structure in pursuit of their "game", the captured photograph.

The extraction and separation of the coal here was also a temporary "visit", another ghostly entity at this breaker. This re-defined coal was largely shipped to outside markets, far away from its home base. But, in this action of separation from the earth, and the further differentiation into various processed forms (chestnut, peat, etc.), does a little (besides coal detritus and dust) of this extraction still remain behind at the breaker, lost in all the processing and reconfigurations? Does this phantom coal presence exist, still manned by invisible hands who still help (as ghostly "breaker boys") in the separation process?

To talk of a "St. Nicholas" at all is potentially misleading, because its location embraces so many "realities" of a continually unfolding, and percolating past. The name, itself, is a cause for confusion, carrying within it conflicting strata of meanings. St. Nicholas (the man and saint) was never "seen" there, and the breaker was not a place of Christmas cheer, or celebratory activity. However, what it replaced was: a community center. In that community building, the inhabitants of the village of Suffolk once did celebrate the Christmas season.

St. Nicholas remains populated today, not by workers, but by these sometime visitors (and most recently, by vandals). The local and area resident, in this scenario, is lost in the accumulated coal dust of yesterdays. Today's descendents of the St. Nicholas workers are at the mercy of these visitor perceptions and their acts of "discovery" (who said St. Nicholas was "lost" in the first place!). Shouldn't this action of "discovery" be reversed? Shouldn't the contemporary "native" be the purveyor of his (or her) own history, and its representation? And is that why the ghosts <u>still</u> remain, making sure <u>their</u> history is told accurately and completely?

Unfortunately, the history of the St. Nicholas Coal Breaker, and its media exposure, is framed in the cultural references of these visitors, who are merely the "ghost writers" of our local history. These visitors become the gatekeepers of this history, assuming the rights and duties of interpreters of the St. Nicholas past, which, in most cases, is both superficial and "ghost-like".

At St. Nicholas, the cultural center of gravity, which is exposed photographically, is sometimes several hundred miles away, a media representation (and revenue-generating enterprises) in New York State, Virginia, West Virginia, Ohio, and elsewhere. This is an act of cultural disempowerment for the people living <u>here</u>, their histories and heritage, who remain "in-absentia" from these representations! These media representations also mask and obscure the struggles and hardships experienced by the workers, whose families may still live here! Where is the local "color" and "flavor" in these visitor photos? Where is the recording of the rich ethnicity and cultural traditions that lie underneath that history (and those photographs of the abandoned structure)?

The end result of these limited visitor encounters ensures that the landscape, at a very "phantom-like" level, reflect a mere flirtation with its overall history. It is a two-dimensional encounter that leaves out the historical, social, and ethnic context that helped to create (and maintain) this structure.

There remains, in the final analysis, only the picturesque structural ruin, a commodity that can bring material rewards to those willing to come here and photograph it. In this sense, the breaker is still in "full operation"! It still generates revenue. And this revenue is still not for the social welfare, and economic security, of the local residents! This is a haunting reality, and all that "remains" of the historic St. Nicholas coal breaker for the local resident. It is another "lost" economic opportunity, and follows the declining path of "coal boom" history. Is this the type of legacy we should be leaving our children and grandchildren?

The multi-dimensionality of the breaker can still be sensed by an "insider". This can be enhanced and embodied in a specific socio-temporal ethnography of the past. In this sensitivity to the past, there is human action, rather than an aesthetic commodity. The St. Nicholas coal breaker can only begin to reveal its true depth when our perspective changes from a haunting uncertainty of its past, to a presence of ghostly certainty and an acknowledgement of that presence.

This shift in sensuous perception requires an active, open, and willing participant-observer, a "being" that is not found in the casual "outside" observer. This perceptual thrust is an entrance into the investigative realm of a ghost excavator. It also means that to be merely physically-present in a place, like these outside visitors, is not a guarantee of unearthing that past drama. Being present does not equal sensing the presence here. The former involves a temporal separation, the viewing of a static past in the present. The latter involves an encounter, engaging the performative actions of that past. This engagement requires an understanding of symmetry, the realization that one (and one's place of residence in all its manifestational forms) is part of the <u>same</u> context. It means that the past continues to unfold in the present. We need to sensually commit to, and absorb, its vibrations. This requires a resident participant-observer, not a "tourist" photo!

A reflecting mirror stands in front of the St. Nicholas coal breaker. It is two-sided, and it contains both haunting and ghostly elements. The mirror allows one to remember the past by seeing ourselves in it today, and in our continuing role to preserve that past. This remembrance and preservation is achieved, if only in fragmentary moments, by learning to see through the eyes of those others that came before us, and those who worked here.

It is the discovery of each other's (past and present) perceptual vocabulary and grammar. It means more – much more – than taking a photograph. A ghost excavation helps us to achieve this perceptual unearthing, by separating the multiple layers of this reflective past at St. Nicholas, and making each one of them relevant, not only for today, but for future generations of area residents.

This type of investigative approach was outlined in a previous book, <u>Ghost Culture</u> (Sabol 2007). I highly recommend it to anyone, both local resident and visitor, who is interested in an engaged performance with the past, be it the St. Nicholas coal breaker, or another haunting drama in a different cultural and/or historical setting.

Finally, it should be emphasized that this engaged excavation is <u>not</u> restricted to haunted locations. It can be used at any local, regional, or national heritage site, one that has significance to the individual (or individuals) that are remembering the past dramas that are contained at these places. These places are part of our (and humanity's) cultural heritage.

B. 2. The Multiple St. Nicholas Fields of Drama

The area that surrounds the St. Nicholas coal breaker has been called by many names. It was Towanamensing, the "land of spirits", the unexplored dense forests of the wandering Delaware hunting bands. The rocks, streams, and animals were accounted for, their underlying meanings inscribed in the forest growth. These were descriptive terms of native spiritualism, containing supernatural elements.

Later, the forest became secularized, and it was replaced by the village of Suffolk, a settlement of immigrants who came here in search of a better life, and who called this land "home". Today, the structure is photographed as the visible remains of the St. Nicholas coal breaker, the largest still in existence in the world.

The tracks that were constructed here for the transportation of the coal cars – to and from the breaker – were called the "stockyards". Cattle were not brought here, however, by range cowboys, though the region, during the days of the "coal rush", was similarly "atmospheric" (the "Wild East"), with its frequent outbreaks of violence and disaster (and its sounds of gunfire, mine explosions, and the "black marias"). This coal "herding" and "drive" was initiated by ethnic miners, not cowboys. The end of this particular "trail" frequently resulted in premature death, and was largely a product of coal dust, accidents, ethnic violence, and the neglect of the coal barons.

In the process of useful (and used) space, the forest became a village, which helped build a breaker, and which, through non-use, became a "ruin", and finally, profitable again for some, a tourist "commodity", exploited in photographic images. In these multiple uses of this landscape, haunting uncertainties certainly were created, and recorded on the structural elements. These multiple and unfolding fields of drama became what Loren Eiseley once described as "not time calendrical".

For Eiseley, time was never a textbook "abstraction", or a definitive and identified moment. So it is here, in this coal region, the time of Towanamensing, the era of the "coal rush", and today's ruins, are one and the same. There are all here still, percolating in the landscape. At night, and in the cold of winter, some themes become dominant over others, and can be sensed more easily, in a "haunted night filled with the turning of vast ears and eyes" (Loren Eiseley).

These ears and eyes are all (still) immigrants here, fugitives assuming the many guises and ghostly masks of many times and eras. These ghosts are free to mingle among the visible structures of this once vibrant, working landscape. But, beneath these multiple layers of ghostly spaces, there are mysteries that still haunt the history traveler, and are still migrating, unresolved, and waiting for answers to their continuing presence:

- What happened to the forest spirits of the Delaware; and
- What happened to those workers in the surrounding breakers the St. Nicholas replaced; and
- What happened to the school that was once located at the present gates to the breaker?

- What became of those students?
- Did they become the ghosts of "breaker boys", who once worked these former breakers?; and
- What became of those displaced Suffolk villagers, and their social center?

Where did they all go? Did they continue to circulate among the replaced additions? Did they continue to migrate about this landscape? Even those of us, who pass this structure today, without giving it a second thought, are we too to be replaced? :

> "Our flesh is linked to these great bones that we recover. Have a care, even in the symbol-shifting brain we may be unable to escape prophetic things. We may be wandering our own way on the roads at night, hear the howl, the guttural laugh of that which will replace us"
>
> Loren Eiseley "Where Did They Go"

As we travel the road that leads through the forests of old, stop, for a moment, at the breaker, and listen for the sounds. They still echo, reverberating, through the night, and the multiple windows of St. Nicholas. Are they the sounds the Delaware followed on their hunts? Are they the instructions given by a forgotten teacher in the Suffolk schoolhouse? Are they the sounds of a village in joyous reunion, perhaps at a Christmas celebration? Or, are they the continuing sounds of the breaker? This auditory complexity should be "broken-down", like the coal before, and separated into their individual past symphonies. Are you willing and able?

Whatever the origin of these sounds (and those other manifestations), these are the "boundary keepers" in this haunted breaker complex. Together, these sensory compositions altered forever this particular space. Let us pause, then, for a moment, be a "breaker" of habitual activities, and listen. If we don't, we may lose these St. Nicholas voices forever:

> "There is a boundary, a boundary between us This is the secret,
> I think of the world, the unseen necessary balance we have
> broken it....we do not hear the boundary keepers.....
> > Loren Eiseley, "The Boundary Keepers"

We <u>must</u> hear (and sense) them. They are bounded to this structure. Their struggles and hardships assigned them this ownership. Remember, these are the ghosts that inhabit this land. Alone, they cannot speak for themselves. They no longer have the physical capacity to do so. They need us to speak for them, by remembering the memories that have left behind for us. We need to vocalize their lives, by refraining from hearing <u>only</u> the silences of a deserted building. We must open our minds, and hearts, to their collective refrain. We owe them that!

We can begin this journey into the St. Nicholas past by shifting our orientation to, and perception of, contemporary space. There is no better location to do this than the "shifting shanty", located on the second floor of the breaker.....

B. 3. Cross(Cultural) Dressing at the "Shifting Shanty":

The Rules of Engagement

Have you ever changed clothes, or your appearance, or even your ordinary, everyday "character". Sure you have! We do this quite often. And in an exaggerated form, we do it every Halloween. We've done it as a child. We do it as adults. But is this exposure of "otherness" merely a Halloween "trick", or something that is not confined to a particular season of the year?

Have you made a costume change at a haunted location, in order to invoke past memories, and stimulate interactive elements from that past? Probably not, right? Have you ever thought of trying it with the "real" possibility of encountering an interactive presence from the past? Probably, not even – for most of you – a "ghost of a chance" for that!

To accomplish this "trick", the rules of engagement are very specific, and you don't necessarily have to find a haunted house! Well, a "traditional" haunted house, that is, to "perform" in. Center stage is nearby. Take, for instance, the St. Nicholas coal breaker. This may be the largest "haunted house" in the world! Or it may be our imagination playing "trick or treat", and that is what may be confusing our sensory perceptions.

In order to determine which scenario is correct, we have to be able to compress (unfold) the past, as it <u>still</u> exists in the present. We have to "re-dress" the reality of that contemporary abandoned setting, and address its past.

What better physical locale to do this than the "shifting shanty" of the St. Nicholas coal breaker, a place where one can "dress for success"! But the rules for ghostly engagement are precise, and should be followed very closely. Otherwise, one does not have that "ghost of a chance" to encounter the ghostly presences!

So, read carefully, and pay attention!! Here are the "tricks":

- Keep an "open mind", and be sensitive to the surroundings and its history, even the history <u>underneath</u> the present structure (its previous occupations);
- Be innocent in approach, almost "child-like". Be alert to the sensuousness that envelopes the visually "drab" environment;
- Understand the spooky "atmospherics": What sensory "cues" could (should) be felt here (what should we smell; what is to be seen; what are the audio manifestations; and, what significance is those insignificant "touches"). This is a quiz: write down your answers, as you experience them! ;
- Do not go alone. This is a matter of security and safety, and just common (not ghostly) sense;
- Record your feelings and observations, all of them, even those that may appear inconsequential;
- Use a "back-up" system of recording. Remember: batteries do "drain" here! ;
- Listen....listen....listen.... (and listen!); and
- DO NOT TRESPASS – GET PERMISSION!

This information will provide the "raw" survey data that you will use to make that "shift" – through performance – to the hidden past that is still wandering here. You cannot connect to that past by being idle, or silent, "waiting" for something to happen. Talk to the entities about <u>their</u> concerns and interests. You will need to write a sensitive, and contextually-sound, "ghost script" – and then perform it! Remember, it is "Halloween Time" here <u>every</u> day, and its manifestations are 24/7!

The performance is the key that unlocks the door to the "shifting shanty". C.A.S.P.E.R. investigators will use these performance-based investigations at St. Nicholas to recall its ghosts. Stay tuned for the results of those performance-based investigations. They will be entered on the C.A.S.P.E.R. web site. This will be our version of the "ghost" channel!

The St. Nicholas "shifting shanty"

C. Seasonal Ghost Excavations

"The sum of our pasts, generation laid over generation, like the slow mold of the seasons, forms the compost of our future. We live off it"

Simon Schama, <u>Landscape and Memory (1995)</u>

The forest is full of "natural" sounds, here in this anthracite symmetrical world. We hear the birds, the water coursing through the valleys, the bending of the trees, the wind traveling through the mine shafts and the empty breakers, all residuals of past voices, and those voices of "unknown" origin. It is the sounds of the four seasons, a perceptual anthracite "surround sound", with the human body acting as the instrument of infusion, absorbing these "ghosts" of the past.

Hunting (ghosts or other creatures), in this dense forest, places a premium on one's hearing, as the main sensory modality of experience. The Delaware knew this, in their detection of distant objects and events. For these hunting parties, the land was "surveyed" with the ear, not the eye. An audible, but invisible, object was <u>entirely</u> "present", though hidden, was very perceptible.

These sounds, and their objective manifesting forms, were especially prominent in the fall and winter seasons, and became the oral narratives of ethnic storytelling ghost stories. This tradition continues today in the "voices" that continue to tell their stories during these times of the year.....

C. 1. The "Fall" of the Anthracite Ghost Story

The inevitable cycle of transformation, where physical "death" (the fall of leaves) is merely the compost for the process of rebirth (after a short, sleepy "silence"), is a spectacular visual delight in these anthracite forests. Do these annual transformations also reflect the processes of ghostly presence here in these woods? Does a ghost lay dormant – their fall season – for a time, before they manifest, again?

If so, what transformations are needed to initiate these processes of "haunting" manifestations? What sensory forms are initially generated? Does the ghostly presence take shape over a period of stages? I believe it is in the fall of the year that the ghosts and phantoms of the future acquire substance and structure. The transformation begins as a haunting uncertainty, and develops into one of certain outcome: a hint that <u>something</u> is forthcoming. The ghost becomes embodied, as a sensory manifesting phenomenon, in the winter snows. And that is why the winter is the season for the telling of ghost stories, a fact Charles Dickens <u>still</u> knows well: he is said to still "haunt" Rochester Castle in England, especially on Christmas Eve, as he takes his ghostly stroll along the castle's fortified walls!

The fall is a good time to walk the wooded areas of the anthracite region, and see these transformations begin to take shape. These woods remain the woods of the "hunt", the search for the Native American and Anthracite ghosts of the region's past. We can "track" (and trek) back to the days of the native hunting parties in the woods once called Towanemensing, if we dare. Do we chance to follow this path back into these woods?

If we choose to enter these woods, where will it eventually lead? What will we encounter along the path during the fall of the year in these woods? There are visions of abandoned structures, perhaps a glimpse of a "bootleg" mine that is still buried in the forest, and that suddenly emerges into sight, with the fall of the leaves. But what else is out there, beyond the security of a well-worn path? The encounter begins when one becomes in-tune to sensory "timing", the period of immersed awareness. And there is no better time for that encounter than in the fall. It is the time when the ghosts begin to re-awaken:

> "I am at last aware
> that there exists
> changelings
> born from a fourth dimension lurking
> somewhere about
> And I am one of these.....
> This means I can see faces
> Where faces are not
> And I know
> a nature still
> as time is still
> beyond the reach of man"
>
> Loren Eiseley, "The Changelings"

It was in the fall of the year that I encountered my "changeling in the woods", described in my book, <u>Ghost Excavator.</u> This was my first encounter with a shape-shifting entity, a prelude to the winter tales of

75

the "Slovak Ghost Children", which, itself, was a postscript to the coal patch ghost stories of Irish immigrant storytellers, narrative delights that foreshadowed the "pleasures" of reality TV! Have we traveled so little perceptively (and in our sensitivity) in the past 150 years? The "reality show" in the anthracite coal region has not changed. It remains a phantom program image of yesterday and today, with continuous "new" shows, beginning in the fall season!

C. 2. Coal Region Winter Tales: Part 1

"There is a single snow which a child stores in his memory....
the first snow that reveals secrets...."
        Loren Eiseley, "The Snowstorm"

My single snow memory was the search for the "white lady". The tale was described in my book, <u>Ghost Excavator.</u> This "investigation" (and "hunt") was the key that opened the door to the coal region past for me. The revelation was quite unexpected, and was an encounter (at night) where:

"The snows that are
older than history,
the woods where the weird shadows
slant....(are found)"
        Robert W. Service

It was in the native woods of the coal region that the search for the white lady ended so long ago, without revealing the details of her secret. But the lesson I learned from that "failure" is <u>still</u> fresh and current in my memory because it brought the symmetry of the coal region landscape into focus: snow, woods, and ghosts, amid the blackened culm banks, and the abandoned mine shafts.

Sometimes, the answer may be a simple black and white response. Winter ghost tales are like that. They begin in the black and darkness of the mines and the woods at night. The ghosts, when they surface, are largely undetectable, in the wintry snows here. That is why we seldom <u>see</u> them. Yet, they are all around us, perhaps 24/7. We must, during the winter months, search for a different sensory means of engaging them, one that is beyond the visual. The answer can be found buried, beneath the accumulations of snow, in those quiet, wintry woods.

To better come to grips with this (and adapt to its "chill"), we will have to remember the woods of this region's virgin past. At that time, detecting presence was auditory, not visual. The Native American hunting bands did not "see" the spirits of the woods. They "heard" them. The key to the ghosts of the past was (is) sound. We hear their "voices". We hear the sounds that no longer have visible meaning. These ghosts continue to echo throughout the landscape, with variations of ethnic sounds and symbolic of traditional values. Do <u>you</u> hear them <u>still</u>? Do we retain the traditional values, <u>still</u>?

These coal region forests, especially in the snows of winter, are full of "natural" voices….in terms of audibility, not visibility. This was the secret tale I learned from the winter snow, and the search for the "white lady". I did not have to see her to sense her "presence". I heard the "voice", echoing in the still of those winter woods. Snow-covered, I was temporally "blinded", but that voice was heard, and it told me to continue to search in those woods….and I did!

These were "mind" (and mine) haunting sounds, the lost ghostly artifacts from the past that continue to circulate here, even today in these wintry snows!

"Kulpmont Hearsay Tales"
By
Craig Czury
from <u>Coalseam: Poems from the Anthracite Region)</u>
(1996)

"I didn't really see this
but I heard it from somebody
who somebody….

A woman was killed one night
tied to a tree and set on fire
on the steep road leading to Minersville.

Broad Mountain around 11:00
and every year on the same date
the same time as her death
you can see her on the hill.
If you're driving around midnight
your car will stall out.
If you go up to her
she will take her with you…."

Such are the winter tales of the anthracite region!

C. 3. The "Ghost" Patches: Winter Tales, Part 2

Winter is the oldest of our personal "ghost" seasons, here in the anthracite coal region because, in that white solitude, our haunting memories were born and continue to grow there. In winter, because of its limiting qualities, we can access and perceive the ghosts in the landscape much more readily. Winter defines the "night country" here. In winter, a miner arose in the darkness, descended into the blackness of the mine, then ascended into the evening, marking the end of his daily cycle of life and work in the dark. In this obscure environment, he heard and envisioned many forms of ghostly presence.

These ghosts are darkened shapes that are created out of the work experiences of the mines. Here the vast recesses of the mind are accessed. They leave traces, as they pass from one coal patch to another, on the surface of this landscape. These traces are all a variation of ethnic red, dipped in mythic tones.

During the winter chill, at a time between supper and sleep, the "children's hour" began. The telling of ghost stories by candlelight passed the hour, as the winter cold darkened the house, as it turned an unearthly black:

> " Those were evenings when, in old houses exposed to snow and icy winds, the great stories, the beautiful legends that men hand down to one another….(began)"
> Henri Bachelin, <u>Le Serviteur</u> (p. 102)

These were stories that were not recounted by women. These were stories, told by the miners, stories about strange noises, and forces, beings, and "signs" in the mine shafts. At night, in these coal patches, the air was filled with the sounds of ghost stories and hauntings that contained a large variety of ethnic flavorings. These ghost stories formed part of the great oral tradition of the anthracite mining region (Korson 1960:302). There were stories of:

- "The dragging chain" (Ibid:304);
- The "headless ghost in white" (Ibid:305);
- "The missing mine mule" of the Suffolk colliery near St. Nicholas (Ibid:308-309); and the
- "Mystery in a bootleg coal hole" (Ibid:310), among others.

As the "coal rush" ended, and the patches became "lost", so did the ghost stories, told during the "children's hour". Each had their own assigned place and time slot in history. The sad part is the thought that these "little deaths" may have been the reason why today, in many local resident memories, the vast majority of the coal patches and their ghost stories, are the "ghosts" of yesterday. Did the coal patch follow the ethnic ghost storytelling by candlelight into the obscurity of another time and life, a haunt dimension?

We do not need a re-birth of the coal patch, but should these ethnic ghost stories that so entertained (and without blood and violence) remain lost in the snow drifts of winter's past? :

"The snow that has no name is just this snow, falling so thick it seems to pause a moment in midair. When I had stared long enough at it, the word that held it showed me only a swirling without a name, a piece of untalkable sky intact above a row of houses, and blankness filling the frames of every doorway, a white that made the dark around it visible"

Larry Levis, <u>The Widening Spell of the Leaves</u>

It was during the winter that only the "shells" of life were active. Activities slowed-down to a different frequency. This was why the fragments of ghostly dramas begin to appear, and take hold. The living and the dead, revealed in these ghost stories, were now "in sync". The frequencies of living and dying were similar, and the miner (and his family), and the ghost, became one, a recurring memory of a once vibrant "coal rush".

The white of the snow, and the darkness of the mines (and mining experience) that surrounded these coal patches, formed a temporary blindness that consumed the house, and its reality. Stories and presence entered through the doorway, and in through the windows, coming from a path back toward the darkness of those mine shafts. The falling snow prepared these coal patch mining families (and us) for what was (is) to come, what still remains unknown. At the same time, there was a preparation for a journey of discovery into that "undiscover'd country" that became the ethnic ghost story:

> "Reality is a very subjective affair…. You can know more and more about one thing, but you can never know everything about one thing: its hopeless. So that we live surrounded by more or less ghostly objects"
>
> Vladimir Nabokov

In these patch homes, neither snow, heat, or the cold were kept out. The only thing that was kept heated inside was the emotional energy created by the ghost story. These homes were "color blind". There were all "painted" a dozen shades of grey, a product of coal dust, soot, and smoke that freely circulated through these patches. In this landscape of the "night country", even the ghosts were grey (at first). That is another reason for the growth of these winter tales. Ghosts became more visible through time (as the "coal rush" continued), and their stories were easily understood. They "blended-in" so well with their surroundings, physical, cultural, and psychological.

Since these company patches were located near the mines, a primary source of ghostly presence, the ghost stories, in these patch homes, were (potentially) endless, both in the imagination, and in the reality of the working conditions of the miners:

> "I wandered over the snow-covered coal site....
> As I paused, I could hear the soft murmur of the ghosts –
> ghosts of collieries, breakers, tipples, culm banks, and of
> miners whose bodies are entombed in long lost mines".
> (Dinterman 1995:99)

The mining of coal in this region has greatly decreased, a livelihood that is almost extinct. But the ghosts, these coal region entities, are still here. They continue to prosper in this landscape, even in the absence of memory and recognition:

> "When at the close of winter's gloomy days,
> our hearts are gladdened by the bright fire's blaze,
> when in each burning coal we seem to see
> fantastic forms of rock and cave and tree".
> Thomas Llewelyn Thomas, "Coal Mine" (1863)

Winter, and the warmth provided by coal, provided an atmosphere conducive to the telling of ethnic ghost stories. In this landscape, the elements of rock (coal), cave (mine), and tree (the surrounding woods) help to "fuel" these stories, and continue their traditional anthracite base of excavations and mining operations. The utilization of this process, in a ghost excavation, will uncover more of these ghost stories, thus providing an "energy" source of emotion and inspiration for seasons and generations to come!

D. A Haunted Town Landscape:
The Excavation of Mahanoy City

I have chosen Mahanoy City, Pennsylvania as a representative example of an anthracite coal region town hauntscape. There are several reasons for this:

- I "grew-up" in this town, and still remember how it <u>was</u>;
- I have experienced numerous haunting uncertainties while living in town (for specific examples, see my book, <u>Ghost Excavator</u>); and finally,
- I still live in the town, and want to record how it is <u>today</u>, with its continuing haunting dramas, and how the certainty of these dramas can affect the future generations of people who will continue to live here. While the town may be slowly "dying", these haunting dramas continue to grow, together with the manifestations of their ghostly presences.

My experiences of life (and death) in Mahanoy City should be viewed in their most general terms, though the origin of these experiences does originate from specific events in my life. What happened here, and what is happening <u>now</u>, are typical manifestations of the events that have occurred throughout this coal region, in its towns and patches, and, most importantly, among its people. My emphasis on Mahanoy City is <u>not</u> a reflection of the events that have occurred there, or their importance in regional history. They were no better

– or worse – than those events, or experiences, that occurred in other local places in this anthracite coal region.

In his book, <u>A River Runs Through It</u>, Norman MacLean states:

> "I am haunted by water. I am haunted by that small mining town. It is the place that helps me feel more at home in this unsettling world, and bedrock to all my writing".

To me, that small mining town was (is) Mahanoy City. And, incidentally, there <u>is</u> a water source that runs through the town! That water source is both a source of my inspiration, and a source for the haunting dramas that still occur there!

D. 1. The Anthracite "Avenue of the Dead"

In the high plateau of Central Mexico, there is an ancient city called Teotihuacan, "the city of the Gods". As an archaeologist, I have walked its avenues, explored its plazas, and climbed its temples many times. I have even sat atop its "Pyramid of the Sun" during a total solar eclipse, and watched ancient ceremonies performed at its base, hundreds of feet below.

Teotihuacan has been called the first urbanized center in the New World. Between the Pyramid of the Sun, and the Pyramid of the Moon, there is a connecting avenue, containing along its parallel sides, various smaller temple structures. This avenue is called the "Avenue of the Dead".

Mahanoy City has its own version. It is called Catawissa Street. Along this street, there are various modern-day "temples", the churches of various denominations (such as the Primitive Methodist, and the Slovak Lutheran) and ethnic faiths (the Irish, St. Canicus, and the Polish, St. Casimir), and a Jewish synagogue. At either end of this anthracite "avenue of the dead" were the schools, the seats of learning for maintaining the ethnic traditions and values of the "old country", and which helped to carry them into the future.

Catawissa Street, like the Teotihuacan "Avenue of the Dead", is now largely an empty space, a ghost of its former self. Most of its "temples" are abandoned, still haunted by the memories of the ceremonies once performed there (baptism, marriage, funerals). Catawissa Street, as an avenue of the dead, emphasizes two important elements of ghost research:

- The importance of understanding the symmetrical dynamics of a contained space, and its various haunting elements. This requires a complete area survey, rather than the "spotty" nature of most "ghost hunting"; and
- The importance of distinguishing between residual and ghostly presence, and the locations where each would manifest.

A ghost excavation unearths the significance and location of these contrasting haunting elements, and their distinction is critical to the understanding of presence and memory. The former is ghostly, while the latter is merely a haunting possibility.

One aspect of ceremony that is still practiced along the "avenue of the dead" is the ritual associated with the burial of the dead. It is appropriate that Catawissa St. also contains a funeral home to prepare the dead, and the living, for burial. There is a European folk tradition that says that if the corpse of the dead individual was not carried along the customary "church road" to the cemetery, it would result in the dead person returning as a ghost. The religious service and the "sacred" route of Catawissa Street were essential elements in a form of "ghost busting".

This street, since it was an avenue for the dead to transport to the cemetery, would <u>not</u> be a place for ghostly presence to <u>remain.</u> The street would be a place of residual memory, a place of passing remembrance of those whose journey had finally ended. These are the dead memories of present-day Catawissa Street, and are similar to those dead memories of ancient Teotihuacan. Can you still sense this residual energy, as you walk along this street? Do the church and school bells still ring to bring the "faithful", the dead back to their former haunts?

REMEMBER:

Churches and schools (see below) can contain a number of anomalous elements: ghostly presence <u>and</u> residual elements. This makes Catawissa Street a <u>true</u> "avenue of the dead! Walk this street, keeping an eye (and ear) open (and "tuned-in") for both types of manifestations!

Present-Day Catawissa Street

D. 2. The "Dead-Line Cemetery"

Mahanoy City is a "natural" ghost town. It is a straight line of continuous streets, with pathways that encourage the passage of ghostly presences. "Ghost paths" have a definite geography, and, on them, one encounters ghostly presence quite often (Devereux 2001:107). In the majority of cases, these "ghost paths" run in a straight line over the mountains, and along the valley floors, like in Mahanoy City. According to Paul Devereux (2001), "straightness was a key association with the passage of the dead, (and) with ghosts" (2001:107). But, the most haunted section of town is the "dead-line cemetery". This is an area of town that no longer exists, because it is haunted by absence, and a diminishing variety of socioeconomic activity. Presence is buried under today's parking lots, and various other "newer" structures (the "hi-rise" building; a food store, and a bank drive-thru, among others).

Where there once was life, and a flurry of activity (especially on Saturday nights), there is now "parking spaces" for both cars, and the old and sick, waiting to join those memories already buried in the past. Do ghosts, and their dead memories, still walk their former haunts? Does Herman Coffey rise from the ashes of his former business to take a night inventory? Due to his poor eyesight, did he miss the "light", and remain at his business?

When I was growing-up in town, these streets were active. There was the Mansion Hotel and Restaurant, "Coney Island", Janowich's Dress Factory, Shandri's News Stand, the Acme, and, of course, the Victoria Theatre, among other business establishments. Do residual energies of these places still echo through the night along these streets? Are there fleeting glimpses of that past manifesting in these newer facilities? Are sounds (and smells) from the past still heard (and tasted)? : the aroma of coffee brewing, a hamburger frying, an old movie scene re-playing, the sound of sewing machines humming, and cash registers "ringing-in" sales and economic prosperity?

Can the people who once lived these experiences (and are still alive) sense today their presence? Should we ask them, before it is too late, and they too become the ghostly memories in this "dead-line cemetery"?

D. 3. The "Other" Mahanoy "Dead-Line"

In <u>Ghost Excavator,</u> I mentioned the "dead-line" of Main St. However, there is another, even longer, "dead-line" in the town. This "dead-line" is today called Market St. Here, there was a "water memory" haunting phenomenon.

Water has always been associated with haunting phenomenon, and ghosts are frequently reported in places where there is a close association with a water source. The British archaeologist, T.C. Lethbridge, author of <u>Ghost and Ghoul</u> (1962), an account of his personal experiences with the supernatural (including first-hand encounters with ghosts), developed a theory that water could record (and retain) strong emotional energy.

The "water memory" "dead-line" was the Mahanoy Creek that traversed the town under Market St. When I was young, this "dead-line" still existed, and it was the basis of my experience (or lack of) with the Slovak ghost children, described in <u>Ghost Excavator.</u> The Slovak ghost children, manifesting close to this water source, were a retained educational residual memory of the schooling that was held in the church.

Today, this section of the water route (and its retained recorded memories) is covered, and forms part of the ground surface of the

Mahanoy Area school complex. Does this mean that the Slovak ghost children, and their memories, have a permanent recess from their studies? Is school <u>finally</u> over for them? And did they graduate into another (after) life? Or, do they still continue to wait for that "hidden" teacher in those 2nd floor classrooms?

The town, itself, may be a potential "fountain" of haunting phenomena, a place where memories were covered and buried. It is only when "cave-ins" occur, such as those caused by the floods of the summer of 2006, that the unearthing of real phantoms occur. Has anyone experienced a dead memory from the past in these excavated flood zones? If you have, please let me know.

Maybe, the era of the ghost storyteller will re-surface, as a re-awakening from the grave of these underground water passages? It is always an education (and an opportunistic delight) to share these types of haunting experiences with others. It keeps the town's ghosts "alive" for future generations. It also preserves the "living" history of this place. If you would like to share your own personal ghost stories, please email me at <u>cuicospirit@hotmail.com</u>. The story origin, and its author, will remain strictly confidential, if you so desire.

D. 4. The "Dead" Houses: The <u>Real</u> Haunts
of Yesterday's Homes

"All great, simple images reveal a psychic state. The house, even more than the landscape, is a psychic state...."

Gaston Bachelard

Mahanoy City is full of "lonely" structures. In fact, there are currently over 400 of them! These are the structures, both residential and business, that are no longer "active" living dwellings. The "breath of life" (or work) are no longer absorbed there. They have become mere abandoned row houses (not homes). Or are they abandoned? Does time still unfold in these houses? Is the presence of the past still active? Does it matter?

I propose that these houses <u>are</u> important. Let us not lose our sense of humanity by abandoning them altogether. There is still potential there. Some of these are <u>homes</u> of the "living dead"! They <u>are</u> homes for the Mahanoy City ghosts. Let's make a distinction very clear, and not a transparent assumption, by defining a mere "phantom" element. I am talking about the difference between a "house" and a "home". The distinction is a reflection of ghostly presence.

A house is physically connected to a geographic landscape point (Pine St. , Mahanoy St.), and is a physical space composed of structural elements (doors, stairs, windows, walls, and rooms). This house becomes a "home" when it is transformed into a physical setting occupied by people. The energy (both emotional and physical) a family invests in the house's physical spaces transforms that house into a haunting field of drama, filled with stories, memories, and experiences that make that house (once merely a physical entity) a "home" (a sociocultural entity). These may be abandoned houses to most of us, but some are still "lived-in". These are the ghost homes of Mahanoy City. The memories of these previous lived moments are the haunting uncertainties that sometimes transform into residual or ghostly certainty.

As you walk past these structures – even in daylight – you might sense, sometimes even smell, the past that hovers over the space. It beckons one to enter, to go beyond the silence of absence, and through the walls of emptiness. There, you will encounter something else....something still heard there:

> "Not a window was broken
> and the paint wasn't peeling,
> not a porch step sagged –
> Yet, there was a feeling.
>
> That beyond the door
> and into the hall
> this was the house of
> no one at all....
> Yet something walked
> along the stair
> something that was
> and wasn't there....
>
> Vic Crume

These homes are still physical entities. They are visible and present. They have spatial dimensions, all too narrow, in these row-house constructions, for a "fulfilling" presence. They may be "airy" structures, through no fault of their own, a product of decay, neglect, and abandonment, and open to the winds of other times, events, and activities. The breath of a multiple past can be heard, as they echo through the rooms and hallways, and freely circulate in the isolated space. On the stairs, movements occur in and out of the symmetrical space. This is the most haunted area of these homes, because it is the most habitually used. The stairs are the liminal spaces between activity and inactivity. It is the space between life and simulated death (sleep). Houses, without stairs, are less likely to become haunted. A ranch house, for example, would lack (but not totally be "free" of) residual recordings. Does that make Barnesville, and the Locust Valley (the location of many ranch homes) less haunted than Mahanoy City, which is the location of multi-story row homes? A ghost survey would provide some answers. The C.A.S.P.E.R. Research Center is investigating this possibility.

In these empty homes, it is perpetual darkness (even in daylight). It is where we recapture and re-contextualize the night. These homes are the places of ghosts. The entrance to this "ghost world" is the door-knob, and it opens more than it closes. It is the ghost pathway, the space of residual presence, and the "key" to these abandoned spaces:

> "I must own that when the door was shut,
> I began to consider myself as too far from
> the living, and somewhat too near the dead".
>
> Sir Walter Scott

We need to excavate these homes, to examine more closely how they have captured the intimacy, the "essence" of their past inhabitants. The ghost excavation is a means to show how the past still engages the present:

"And the old house
I feel its russet warmth
comes from the senses to the mind"
Jean Wahl, Poemes, p. 23)

The unearthing of this ghostly presence requires a sensuous excavation, a tool kit of sight, sound, smell, touch, and taste for the past, a particular Mahanoy past. This sensitivity involves a highly-charged emotional scholarship. It requires a "deep dig" to expose the ghosts within these structures. It requires the personal reflexivity of a ghost excavator.

A ghost excavator extracts the drama, and its principal actors, from these homes. We need (must) open the silent portals of these structures, and ask ourselves: Does the excavation make these homes useful (significant) again? If not, should they permanently be "laid to rest".

These homes, that were once lost, have the potential to live again, without us physically inhabiting them. They are the respite and retreat of collective anthracite memories. Our excavation can re-animate the past lives and drama that are recorded in these structures. As Bachelard has said:

"It is as though something fluid had collected
our memories and we ourselves were dissolved
in this fluid of the past"
(1969:57)

The result of all this home unearthing and extraction is that "the entire reality of memory becomes spectral" (Bachelard 1969:58). <u>We</u> have made the abandoned home "haunted" through our excavation process. Yet, the ghosts came from within, the house itself and ourselves:

> "A house where I go alone calling
> a name that silence and the walls give back to me
> a strange house contained in my voice
> inhabited by the wind.....
> Pierre Seghers, <u>Le Domaine Public</u> (p. 70)

These are still breathing homes, homes of wind, dust, crawling creatures, and ghostly voices that hover in a transitional state between physical reality and haunting potential.

So, the next time you pass by and see one of these homes in Mahanoy City – and there are many – be a little respectful. Do not show a disdainful look. Don't be deceived by outward appearances. Remember, the life that was lived there before they were abandoned by the "living". Remember the heritage of ethnic tradition that came before and once lived there.....that still may live there. And perhaps, with permission (REMEMBER: DO NOT TRESPASS!!!), you can tap into that past world.

Show a little kindness, and use a gentle approach, by re-calling (and remembering) the home's history. For it <u>still</u> lives, in one form or another, through its haunting presence. The experience you will sense here will not be a frightening one. On the contrary, it is a learning one.

You, dear reader, should be a little proud. It was once a home, after all, and not just another row house. People lived there.....and their memories (good and bad) can still be sensed by those willing to do a little "dirty work" through the "excavation" of these abandoned structures. For these unoccupied homes are the "living museums" of Mahanoy City!

Finally, remember that each of these "empty" structures represents another fragment of the gradual and slow death of a sense of the Mahanoy City community, and another loss to some past (and meaningful) social interaction and relationship. Though these structures are apparently physically and visually abandoned, remember:

> "However empty your neighborhood may appear,
> at least in a haunted house,
> there's <u>always</u> somebody home"
>
> (Skal 2002: 121)

Photo of a now "Abandoned" Mahanoy City "Home"

A Ghost Survey:

Who once lived here?

What is the geneology of habitation?

What events and activities occurred in these living spaces?

What manifestations <u>still</u> occur here?

Is this a "home" for ghosts??

D. 5. The Personal "Ghost Homes"

There are houses that I grew-up in, in Mahanoy City. They speak of a time that is now "out of sync", in which the rhythm of activity differs from today. It is a faster pace now, so their haunting presence is not easily distinguishable. They are still physically there, these three homes, though they might have changed their external appearances, through personal changes in taste. Inside each, I remain, the activities and experiences contained and recorded. Though physically absent, I am that haunting uncertainty that remains.

The house that is located on West Pine St., I no longer can recall the interior memories. This was my first home. The second, located on West Market St., I can remember only the indoor partition, which separated the "living" area from the "dining" area. Such arbitrary distinctions, in these narrow row homes, were unwarranted, for they served the same functional space. The third home, also located on West Pine St., I remember well. There are many memories still stored there, hidden from the current occupants.

All of these homes contain fragments of my childhood and early youth. These are places where even the words, though no longer spoken by my lips there, are not lost with the passage of time. They remain imprinted in these homes, waiting to be heard again by those willing to listen. Surely, they will remain unspoken to the new residents (at least for a time), as they are themselves building (and depositing) their own haunting vocabulary and grammatical structure. The time there now is not my time. So, my memories remain there, as a series of haunting echoes from the past that may be too personal to be understood by the current owners.

You may ask, do I wish to go back there, and visit these houses, and recall the "home" I lived and experienced there? If I could, would I want to re-live <u>that</u> past? Loren Eiseley answered that question many years ago:

> "Can there not be miniature time? Some place where one stays
> forever at the kitchen table,
> on the same page of one's book,
> with one's parents looking on....
>
> I do not recognize this alien, grown-up body....
> I am there, there, in the yellow light in the kitchen,
> reading.....
> We are all there. I did not grow up....
> I have rushed like a moth through time
> toward the light in the kitchen.
> I am safe now. I never grew up.....
> "The Mist on the Mountain"

I remained in Mahanoy City – even when I left. I am there, sitting alone, in all those homes of yesterday, knowing tomorrow I will be elsewhere, roaming the coal region landscape anew. I have no choice. I am a ghost excavator. I have returned to the haunts of my youth. They are one and the same, this past and present (and future). The mines of my childhood have become the memories of my adult excavations, as I unearth these memories, back....to the present. This <u>is</u> a ghost excavation, is it not? This is not the simple disinterment of local ghost stories and lore. It is more – much more! It is the unearthing of a personal past that is seldom exposed (or even perceived) today by most local residents: do <u>you</u> recall the homes of your youth, and the haunting memories that are contained within them?

D. 6. Mahanoy City's "Stairway to Heaven"…
Up to a Continuing Presence?

The present Mahanoy Area Middle School building was once a stairway to a better life through education. Today, only the ghosts of former dreamers and their memories travel up and down these stairs, along with some health-conscious types (including myself). Don't they (I) know that is only prolonging the inevitable! Give us an "A"….for apparition. We all may soon potentially become one, as we continue to climb that stairway. Our residual energies form part of, and join, those of countless others who once climbed those stairs. Those that walked those interior halls (including myself, and my daughter) are part of the symmetry of this coal region landscape:

> "Oh the spirits still are yearning
> for those hallowed halls of learning
> where they spent so many hours
> in the lost lost days of yore.
> They come in wraithlike forms….
> and walk the halls to classrooms
> as they often did before".
> Docia Williams, Phantoms of the Plains

The Mahanoy Township school stairway was a symbol of the continuing presence of a large student body. Today's school enrollment is merely a "phantom" of what it once was. The enrollment in 1932, at its height, was 642, with a faculty staff of 27.

There should be some manifestations of a ghostly curriculum, and other non-academic activities, at the school. This potentiality may become a haunting reality, as the current building is being re-modeled. The introduction of new window panes at the school should help to "trap" any residuals that are contained there.

Are there any interactive entities still willing to continue learning at the school? There is an urban legend there of a "hunchback" and his dog. Both are said to prowl the basement area of the building. Have the present-day workers experienced anything? What really is going on in that basement? A ghost excavation might unearth, and "flesh-out" reality from legend?

Maybe, the stairway should be used again, to instruct future generations about the importance of history, heritage, ethnic tradition, and education. Is the "hunchback", a symbol of that desperate need?

The Mahanoy Area Middle School "Stairway to Heaven"

D. 7. The "Lost" Schools: Ghostly Residuals
Still Waiting to Instruct

At one time, the Mahanoy Area had a large number of schools, both parochial and public. In 1900, the first public high school in the Mahanoy Area was built in Coles Patch. It was an immense structure, surrounded by trees and eager students. Where is that school now, and what became of this building (or its remains)? What happened to that eager craving for study and learning? Today's contemporary students are mere "ghosts" of those highly motivated students from the past.

Where are the one room school houses in Morea, New Boston, and the Vulcan? What happened to the books and school supplies? What became of the old Suffolk school house that was replaced by the St. Nicholas coal breaker entrance gateway? Are "lost" instructions still being repeated there to a now empty and unoccupied space, where desks were once situated?

What about the St. Nicholas elementary school, once located across Route 54 from the breaker? In the absence of a surface structure, is there something still remaining of the "schooling" that was taught there? Surely, there should be some auditory manifestations that still remain? The constant noise from the breaking of coal across the road must have influenced the classroom instruction? Does that noise still remain? And, if it does, are some children still waiting to be heard?

Or is their silence a residual remembrance of not being heard through the once active breaker noise?

One loss is certain. The Slovak ghost children have graduated. Their diploma was the covering of the Mahanoy creek in that area. The uncovered flowing creek was a source of their continued manifestations, a "water memory" of their willingness to learn, even after death. Now that the creek is covered, the Slovak ghost children have moved on, to seek their fortunes in the world beyond the confines of the school they once knew so well.

It seems appropriate that their school days were replaced by newer ones that form part of the Mahanoy Area School District (and not just Mahanoy City, or even Slovak culture). Today, there is basically one large educational complex. One loss is another gain. Educational energy (and its creativity) does not dissipate. It merely changes form, and leads to newer (and fresher) avenues of instruction, exploration, and discovery.

Is there anything that remains of the Polish school beyond the present empty space on Catawissa Street. Hopefully, at the very least, the Polish ethnic tradition and values still are followed. The demolition of a mere building should not mean a loss of the customs that had survived for generations, and had traversed an ocean. That loss would be hauntingly tragic.

Did all these scholastic buildings just disappear from the Mahanoy Area's curriculum, to become habitations for other memories? These schools, no matter what size or level of instruction, were (are) important sources for residual (and habitual) manifestations from the past. This is because these instructional spaces, that once occupied the student's minds, continue to percolate with the sounds and movements of learning.

These "lost" schools are ghostly reminders that thoughts, no matter how fleeting, do not "die". They continue on (if recalled) endlessly into the future. That is why these locations are good sources of more than ghostly or haunting presences. They are locations of past inspiration that can help educate future generations about the history of the Mahanoy Area, <u>and</u> serve as a framework to rework old ideas, and create new configurations of thought and action.

This is an educational heritage that goes beyond labeling these structures, where they still exist, as possible "haunted" locations, merely because of their contemporary "spooky" appearance. That is why the C.A.S.P.E.R. Research Center is both an investigative, as well as an educational, center. We excavate these structures in order to educate the general public about their significance and value to future generations who continue to live in this Mahanoy Area.

The St. Nicholas Elementary School

The Contemporary Remains of the School

D. 8. Memories of the Haunting Inferno(s)

"Pueblo chico, infierno grande"
> (small town, big fire)
> Old Mexican Saying

Mahanoy City has had its share of big fires, and these fires have produced some disturbing images, even perhaps some apparitional forms. Row-house construction, even in the "business district", has flamed these fires, creating not only haunting images, but ghosts as well. These are the "ghosts" of a forgotten geography of the town. What once was, is no more, and has become something forgotten in memory. There are also episodes of interruption, the displacements of family and social relationships. These fires burned more than houses and personal belongings. They ended the accumulation of memories at particular locations. Such was the case for the blaze that occurred on June 15, 1985. The fire destroyed the St. Nicholas Elementary School Building. It was an empty building to most people. But it still contained the "spirit" of instruction. That was the greatest loss the flames consumed. These fires did not claim as many lives, as they did personal and familiar remembrances, supplanting the old memories with new painful visions of decay and absence.

To me, the most personal fire experience was the one that occurred on December 29, 1958, which destroyed the buildings where the parking lot next to the Market is now located on Main St. In that fire, Herman Coffee died. I remember the stories that circulated about his death, and how he was found. If ever there was a case for a haunting, this was it!

We would expect to see "familiar faces" coming out of the smoke in these fires, the friends and neighbors who inhabit these engulfed homes and/or businesses. Yet, this is not always the case. In Centralia, for example, there have been reports of apparitions of miners (unknown faces), wearing their mining helmets, walking out of the smoke, near the cemeteries of the town. These are momentary visions, since these miners would suddenly disappear. Were these the disturbed souls of those buried in the mines? Do these ghostly miners still work in the mines around Centralia? Whoever they are, they are not the contemporary residents of the "ghost" town of Centralia!

Marie Davis Williams has written a story called "Anthracite" (contained in Philadelphia Stories found at www.philadelphiastories. org). In this article, she tells a story about the following:

> "A miner tells a priest that the mines are damp and cold and unlike any picture of hell he has ever seen in his prayer book. The priest makes the sign of the cross over the man and whispers to the miner saying, 'You haven't dug deep enough yet".

At Centralia, the depth apparently was perfect!

Do these fires reverse the polarity of time and space? Is presence no longer the existence of contemporary residence? Does this free movement between the past and the present also occur in these Mahanoy City fires? If this is true, are these the ghosts of former miners (and/or their families), or the ghosts of more contemporary residents? That is one question I ask myself when these fires claim another torched victim, who relives a highly personal, and agonizing, experience of hell on earth!

These fires are a legacy of the Mahanoy City past, and, unfortunately, it will still be a major concern in its future. In town, this particular past will not be extinguished alone by the "water memories" of the local fire companies, no matter how proficient they are! They will remain alive, an unfortunate probability for this row-house construction, and is a fate worse than death, because it may not be individually controlled. Is this why the ghosts remain in those houses that once "experienced" smoke and fire damage, to continue the tradition of the "haunting inferno"?

These fires traverse the ethnicity of every neighborhood in town, irregardless of the original "dead lines". This is one ghostly legacy from the past that needs to be extinguished! These fires are a palimpsest of erasures, and a beckoning re-call to the ghosts who <u>still</u> inhabit this town, and these row-houses!

The Coal Region Hauntscape Excavation Summary

A. The Coal Region "Stairway to Heaven": A Missed Opportunity, or a Reason To (Still) Remain?

The opening lyrics to the classic Led Zeppelin song, "Stairway to Heaven", echo the haunting uncertainties of this anthracite landscape. A major question, though, <u>still</u> remains unanswered: Why do these ghosts remain in this coal region landscape? Have they missed opportunities, through no fault of their own, to "enter the light"? Or, have they chosen to stay for other, more personal, reasons….? :

> "There's a feeling I get when I look to the West
> and my spirit is crying for leaving.
> In my thoughts I have seen rings of smoke through the trees
> and the voices of those who stand looking….
> And a new day will dawn for those who stand long
> and the forest will echo with laughter".

Do the ghosts stay because this is <u>still</u> the land of Towanemensing, the "land of spirits"? Is it because it is located to the West, the heart of coal country? Do they feel this is <u>really</u> "home", after all those migrations? Those rings of smoke, and the laughter and voices in the forest, are what the bands of Delaware heard on their hunting forays into this forest, as I too heard them. They were my "changeling in the woods", described in <u>Ghost Excavator.</u>

The symmetry of the land remains, even today in this technocratic society. We are <u>all</u> immigrants in this land of spirits!

> "My dad told me that when he was young
> he used to go swimming in a stripping pit
> until a kid went swimming there at night
> and drowned.
> My dad said everytime he went back
> he heard the kid screaming and splashing
> in the water...."
>
> Craig Czury, "Kulpmont Hearsay Tales"

B. The Aquatic Memories of the "Stripping Holes"

In the coal region, swimming in an old strip mine was a common practice. It was both exciting and deadly. I personally know of one death attributed to swimming in these stripping holes. This dangerous pastime, of the past, has resulted in their haunting uncertainties.

Today, most of the places associated with these haunts are (thankfully) buried, and potential ghosts are now trapped underground. Or are they? Are they searching for a release? Have they developed an exit strategy?

New structures dot the landscape, covering the former mining operations. One of these includes parts of the Mahanoy Area football stadium, and the adjoining tennis courts. Can the voices of the dead still be heard beneath the shouts of fans from touchdown runs at the stadium, or balls hitting a tennis racket? What happens when there is no game? Do the voices speak to silent spaces, and to empty seats? Do these voices even exist?

What happens at the bottom of the former "swimming holes". Remember, water may have a memory. Does this memory travel along underground shafts, once the work sites of ethnic miners? Do the numerous "bootleg" mines in the area release these "trapped" ghosts? Is this cause and effect, or just plain myth-building, or part of folklore tradition?

There is a local legend that miners are said to haunt the tunnels of a huge underground mine city that stretches, some say, from Pottsville, in Schuylkill County, to the Scranton-Wilkes-Barre area, in Luzerne and Lackawanna counties. It is said that the underground paths that the ghost miners follow to communicate with each other are avenues of transport from their "hell" on earth. Is that why "ghost" miners appear, above ground, in the smoke of the fires burning in Centralia? Were they seeking the "light", and thought Centralia was the entrance to it? :

> "The band of miners leaves the morning light,
> and soon is shrouded in the shades of night;
> Some to the gloomy mine's remotest end,
> to scenes of daily toil their footsteps bend,
> their labor o'er – they turn to seek in vain
> the path that leads them to the light again...."
>> Thomas Llewelyn Thomas "Coal Miners" (1863)

Some of these ideas of ghost miners and their underground passages do resonate with other similar settings. For example, in the coal mining region of southern Wales, the ghosts of dead miners killed in the mines are said to rise from the ventilator shafts, and are known as the "spirits of deep navigation". Do the aquatic memories of the "stripping holes" support travel along the mine shafts, becoming, at the surface, these "spirits of deep navigation"? Are the Centralia "ghost" miners an American version of the Welsh "spirits of deep navigation"?

Whatever they are called, the beings which originated work or play on the surface, eventually ended there, in the mine shafts; and, after traversing the excavated earth, came back "home" again. If this is true, it is an appropriate finish to a continuous cycle of life and death in this symmetrical coal region landscape. The connected chain of living and dying remains unbroken. It is a circle in a series of transformations, and water may be the transmitting agent of both origin (life) and final product (death), in the form of ghostly presences.

This is the anthracite region, a setting for potential change, and a repetition of past events and experiences, and all of this, the product of a haunted landscape. One day, perhaps at physical death, all of these ghostly presences will scatter and inhabit the locales of our fondest memories. These are, after all, our residual experiences. This is not about something that is paranormal or supernatural. It is about the building of memories, albeit haunted ones, piece by piece, as if we were creating a new living (and expanded) version of ourselves.

This does not make us creators of our own Frankenstein monster. On the contrary, we are experiencing the world, and as we do, we get "fatter" (from the memory and experience), as we continue to grow older. Is physical death, then, the final "bursting of the seams"? Is that why reports of hauntings are usually fragmented, consisting of perceived "bits and pieces" of individual personalities, emotional dramas, and habitual/mundane activities?

Is this insight, the last phase of the current excavation of a haunted coal region landscape, or merely a pause in a recurring ghost season? A final (?) chapter will attempt to answer that question, albeit, not completely to everyone's satisfaction, not even for the ghosts themselves!

The question will be rephrased, in that final chapter, to the following:

- Is this the end of <u>the</u> ghost excavation season; or
- Is it merely <u>a</u> excavation season, one in a continued exploration of the ghost and haunting phenomenon in this anthracite region?

The End of the Excavation Season

"The living are the dead on holiday"

Maurice Maeterlinck

"The splendours of the firmament of time
may be eclipsed, but are extinguished not; ....
The dead live there and move like
winds of light on dark and stormy air"

Percy Bysshe Shelley

Human life is so short-lived, but ghostly memories are timeless. They continue to haunt us, well beyond the grave. Within each of us, you and I, there is a key to a door that unlocks various ages. The passage through that door is our ability to see in a mirror, and remember, back....then. It is the "unforgetting" time, when the lessons of the past are reflected back to us. They stay with us, even after death.

We humans have always had two ways of observing. We see physical reality, yet believe in an unseen presence. In the coal region, unnoticed phantoms can be sensed in the woods, and can be felt in the mines. A scent of ghostly presence lingers at both locations. A ghost excavator has the ability to unearth both these worlds: the surface vision (however unexploited and unnoticed it is today), and its underlying and unseen spirit (despite its fragmented and fleeting moments of manifestational forms).

Both visions are necessary and vital to us. But, a reliance and dependence, on one or on the other, may prevent new and fresh ideas from surfacing, or worse, not allowing the impact and significance of the past from being fully understood and appreciated. This "spirit(ed) unearthing", and its surface presence, is a digging process that is critical to life, and the understanding of death. Without it, that sense of awe and mystery, which has been an essential part of humanity since the recording of history, would soon vanish. Without the ability (and desire) to dig into, and out of, our personal histories, we would lose our humanness. This makes the presence of this past in contemporary life, a true haunting memory, and not a supernatural or paranormal event.

By applying a symmetrical vision to our sensory perceptions, the past and contemporary worlds retain their "naturalness", a sensitivity that has almost been completely lost in this technocratic age. The short-lived dramas of a human life, and its "living" memory, can be extended into a future age, through the excavation of these personal haunting realities.

More than 150 years ago, back several generations, the Danish philosopher, Soren Kierkegaard, observed that maturity involves "a critical moment where everything is reversed, after which the point becomes to understand more and more that there is something which cannot be understood". This is surely the mind set of a ghost excavator. I "see" the relevance of these statements in my own life. I have come full circle, from a hunter of ghosts in my youth, to an excavator of personal haunting memories in my mature years. What I have attained, at this stage of my life, is the realization that I am not a seeker, but rather the prey, and the object of my search. I am that "ghost within" all of us!

In the words of Blaise Pascal, French physicist and philosopher, "You would not seek me had you not found me". My life has been a journey back to an inner sanctuary of coal region memories. In that journey, I had not fully grasped what I was really seeking. I became aware of it through the unearthing of the (my) past that was contained within my own ghostly memories, a presence that was there all the time!

Today, I stand at a point where the unknown not only comes into being, it becomes "natural". This being, through my excavation process, stands close to the doorway to the past that lies partially hidden, a world that is still beyond my full comprehension. This does not mean that it doesn't exist in reality. What it personifies is that the true meaning of vision still lies buried within me. It remains only partially uncovered.

The understanding of this provides me with a key to that door, and a passage into a realm where natural and supernatural no longer exist. It is a place that does not distinguish them as opposites. What they are, in our liminal state of living, are two sides of one's nature, as we live out our lives on a daily basis.

I am, at last, face to face, with my natural self, one encompassing both visions: it is the reality of a life "lived", and the spirit that is contained in that "living". Loren Eiseley once said,

> "Perhaps there may come to us then,
> in some such moment, a ghostly sense
> that an invisible doorway has been
> opened – a doorway, which, widening out,
> will take man beyond the nature that he knows"
>
> (1972:181)

Is this, then, the light at the end of a tunnel so talked about in near-death experiences? Must we physically die to discover the true nature of the world? Do we become, finally, at the end of that tunnel, the ghost within.... ourselves?

Finally, has the end of our excavation season – physical death – revealed what we already knew? Or, is there another season, where the unearthing of the drama in the field continues....? Only our spirit will know that for certain....because that is a true vision of our continuing existence. In the final analysis, the end of an excavation season is merely the beginning of a new one!!

Appendix

A. The "Real" Meaning of the Spirits of the Coal Region

This book has applied an investigative philosophy that was centered on the ordinary and mundane, as indicators of possible haunting phenomenon. This phenomenon occurs in "lived-in" coal region houses, where un-eventful (and non-dramatic) experiences occur; and in abandoned coal mining structures, which may still retain residual elements of the "coal rush" past.

I have used direct story-telling, a ghostly version of oral history that was characteristic of the period of the "coal rush" ethnic storytellers. These stories and essays are non-linear narratives, containing fragmented episodes that are symmetrical in nature, and characterized by their haunting portrayal of a forgotten past. These portrayals involve individual (and personal) actions and decisions.

It is the materiality of the coal region past that has continued to materialize in this landscape.

These narratives are also an interpretation of "dark spaces", a "night country", seen only with the limited illumination of a coal miner's lamp. Yet, these dark spaces are not confined to the underground passages, located deep in the mines. Darkness occurs (and quite often) on the surface of the associated mining communities. These habitation sites are "patched" to a particular destiny that is associated with the mining of coal.

We, as ghost excavators, can see through the darkness. Barriers, once fortified by indifferent coal baron/owners, and uncaring coal companies, now disappear, as time unfolds unto itself. The archaeological use of these "ruins", and their contemporary disuse, now serve a more complete and representative purpose. They can brighten this darkened landscape, as "dead" and shadowy shapes and forms begin to solidify, and materialize in contemporary space.

Ghost excavation is an effective entry-point into a regional history that is populated by a rich, and diverse, ethnic tradition. The "out of time" (and out of sync) coal region landscape can become "mainstream" (and significant) again, through the popularization of its ghostly and haunting landscapes. Coal, as the centerpiece of this anthracite region, has become useful again, a cross-generating information source. It can also "ignite" a fire of renewed energy, and a "heated" discussion on the extent, intensity, and unique manifestational aspects of the anthracite haunting presence.

One reason, albeit a major one, why ghostly presence has not readily been detected previously in the coal region, was (is) due to the high consumption of alcoholic spirits. These types of "spirits" dull the senses, not stimulate them. Because of this, the ghosts remain largely "unseen", a "no nonsense" presence, and are mere "flights" of fantasy and imagination. Until now, these presences remain the source of ghost stories, not substantive past coal region haunting dramas!

Isn't it time to end this "masquerade", an obvious Halloween "prank", and start to expose the sensual "reality" of this coal region, and without the help from those other "spirits"? We can begin by mining the coal again….and the ghosts that were born here during the "coal rush".

Let us "toast" their continued presence here, as we begin the new "coal rush" to our region, by a wave of new (and continuing the rich tradition of the region), temporary migration of peoples, the "tourists". For these "ghosts", long absent from our minds and imaginations, are now the "native" residents of this region, and are not the immigrants of their past!

B. Making Sense of the Anthracite Ghost!

In doing research, and conducting field investigations, it is important to focus on <u>all</u> the multiple sensory dimensions, characteristic of <u>any</u> given landscape. To me, this emphasis on sensory holism is even more critical in one's perception of the anthracite coal region, its history, the diversity of its culture, and the people who came to settle here. Landscapes are not just 'views' of a place. They are intimate encounters. They are not about seeing the land as it is, but about experiencing it with all the senses we have. I would include that 6th sense of encountering and perceiving the "coal rush" past that still permeates this region, and remains largely invisible to the naked eye.

This multi-sensory experience is the reality that is missing from a visitor's perception of this environment. A phantom image of the coal region exists in these tourist's minds. These are "ghosts" of misunderstanding and biased interpretations made by those "blinded" by a subjective vision of existing (and abandoned) structures and sociocultural institutions.

The real ghosts, those that continue to interact in this environment, lie <u>underneath</u> these <u>surface</u> elements of physical and cultural features. A significant and material anthracite culture is a "ghost culture" to many. Yet, there <u>is</u> a "ghost culture", one that gives expression to a particular set of sensual relations. Making sense of this "ghost culture" involves a particular way of sensing, not seeing. This is an active process, and involves seeing, hearing, touching, smelling, etc. as primary sensitivity tools. The distinction between subject (living human beings) and object (anomalous manifestations) has no sense or purpose here.

Meanings are mediated through ghost excavation procedures, and an investigative performance. In this re-contextualization of coal region haunt dramas, more attention is focused on:

- The intensity and duration of each sensory modality that is perceived, measured, and recorded;
- The sequencing of anomalous manifestations and their "sensing", as they unfold in physical space; and
- The selection channels that are utilized and that focus on particular individuals, as socially-significant persona, in which communication is initiated.

This communication process is intersensorial, and is a key factor in re-generating (and re-establishing) a new "coal rush". This "coal rush" is an exploration of multiple and symmetrical sensory relationships involving embodied experiences. These intersensorial experiences can replace the sterile anthracite coal region image, making sense of these "ghosts" as full-bodied images. What was once a "phantom" perception, or a local ghost story, is now a well-developed haunting drama that is not so easily lost or forgotten again!

These local manifestations can become universal, emotional stories for ages to come. Doesn't that make more sense than a view of absence, with its vision of empty and abandoned structures, and an image of visitor-perceived coal region "ghost" towns?

C. Anthracite Ghost Hunting: A Call for Mobilization

The anthracite coal region, as a cultural (and industrial) heritage tourism destination, is a specific geographic entity, consisting of various dominant physical entities and features (breakers, culm banks, collieries, "patches"). This book defined a means to transform this space into something that is hauntingly significant. The means to this significance, I suggested, is through the regional (and ethnic) ghost story.

Coal region heritage space consists of geographical, socio-cultural, and emotionally-laden landscapes. Its geographical component contains a main narrative that consists of a sense of loss, and a future economic uncertainty. This narrative can be readily observed in this region, and is contained in the towns, houses, coal patches, support structure, and other "work spaces" that are located in this landscape. The socio-cultural and emotional landscapes consist of the personal (and individually-unique) stories, memories, experiences, and "home" spaces of those inhabitants who form the core of the main narrative.

There is potential to dissolve these boundaries – across space and through time – through the dual processes of symmetrical space and unfolding time. This process is "unearthed" through the excavation of the main narrative, separating it into its component ghost stories.

This unearthing was initially begun in <u>Ghost Excavator</u> (Sabol 2007). In that book, the physical geography of the mines, town, and woods translated into the stories of the "white lady", "the Slovak ghost children", and "the changeling in the woods" (respectively). Those ghost stories were meant to cut across temporal barriers, erasing the obstacles that separated "their time" and "now". They were also metaphors for more universal themes of a liminal reality that also crossed physical and cultural (ethnic) boundaries.

The idea behind <u>Ghost Excavator</u> was to present local ghost encounters within the context of a personal memory of various coal region landscape settings. To experience these coal region spaces, on an individual basis, involved an exchange, a "dialogue" with the "something else" the physical space had contained, but was now temporally absent from. Presence had declined in significance, and its importance became lost.

The purpose of the present work was to address the continuing process of unearthing and translating "lived" memories and experiences into a form of theatrical performance that could serve the needs of both performers (the interested public, including ghost investigators) and the "audience" (the anthracite ghosts).

Ghost stories, and their scripted ethnic performances, acted as a catalyst in this symmetrical anthracite space to unfold time, by providing "a legitimate theatre for practical actions" (De Certeau 1984:125). These "practical actions", I suggest, are the excavation rehearsals of field investigative activity. This is a creation process that provides a framework for cross-cultural communication. Here, there is a notion of performance as a rehearsal (and review) of <u>a</u>

life, residually present (and temporarily lost). Through the rehearsal performance, memory is transformed from a "ghost story" to a real haunting drama.

The rehearsal triggers ghostly memories. Initially, however, the haunting drama, though manifesting in physical space, went unnoticed, a story waiting to unfold and be told anew. The "ghost story" remained suspended, as a forgotten memory. This is the "ghost within" concept, a haunting drama in the field, awaiting its "storyteller".

Engaging people in a ghost excavation, as part of an anthracite heritage tour package, facilitates the act of experiencing the "ghost within", which is reflected and represented in a personal story that is not a part of that individual's (as a visitor) contemporary storyline.

This coal region ghost tour/excavation eliminates the rigidness that is perceived of fixed geographic boundaries, fixed realities of presence, and biased images and perceptions of this landscape as a declining, static, and insignificant environment. Rather, this region begins to percolate with the ghosts of its past, a resource in sufficient supply. It "uplifts" the "spirit" of the local economy. But, remember, these ghosts are not material objects, subject to "commodification" and exploitation. They are surviving personalities of human life. Thus, a ghost excavation/tour also means the unearthing of the human reasons for their continued presence. The tour becomes a rescue mining operation, as well. We need to listen to their story, and then, we need to help them.

The idea of memory exchange through excavation rehearsal, one that cuts across physical boundaries and between different cultural traditions, is a means to end the rigid (and "dead") geo-cultural boundaries of an "in-situ" anthracite coal region culture. Paul Ricoeur, in the book, <u>Paul Ricoeur: The Hermeneutics of Action</u>, comments on the concept of memory exchange:

> "The liberation of (the) unfulfilled future of the past
> is the major benefit that we can expect from the
> crossing of memories and the exchange of narratives".
>
> (p.8)

This memory exchange is theatrical, because it is performative. The rehearsal reveals a "true" story, however fragmented it may be. These rehearsals are also an exchange of images between different cultural traditions, the "target" objects, which are used to identify contextual (and individual) significance. These exchanges are <u>not</u> necessarily verbal responses, such as typical EVP recordings in standard ghost field investigations. They are a broader, more encompassing, investigative activity, one that involves the exchange of ghost culture elements, telepathically-communicated through <u>non-verbal</u> communication. Isn't this what telepathic communication is? Isn't that how a "true" ghost communicates?

I know you can "hear" them calling you! Their presence has always "spoken" for them! What are you waiting for....a verbal or written invitation? Let us begin the excavation <u>now</u>, before it is too late to respond to their "voices"!

Bibliography

Bachelard, Gaston. The Poetics of Space. Beacon Press: Boston.
1969

Bakhurst, David   "Social Memory in Soviet Thought" in D.
Middleton and D. Edwards (Editors) Collective Remembering
Sage: London. pp. 203-226. 1990

Blomain, Karen (Editor).   Coalseam: Poems from the Anthracite
Region. University Of Scranton Press: Scranton.   1996.

Bartoletti, Susan Campbell.   Growing Up in Coal Country.
Houghton Mifflin Co: Boston.   1996.

Bruchez, Margaret Sabom.   "Artifacts that Speak for Themselves:
Sounds Underfoot In Mesoamerica". Journal of Anthropological
Archaeology. Vol. 26(1): 47-64. 2007.

DeCerteau, Michel.   The Practice of Everyday Life. University of
California Press: Berkley.

Devereux, Paul. Haunted Land. The Bath Press: Bath, England.
2001.

Dinterman, Walter L. Anthracite Ghosts. University of Scranton
Press: Chicago. 1995.

Eiseley, Loren.   The Firmament of Time.   Atheneum Publications: New York. 1969.

The Night Country.   MacMillan Publishing Company: New York. 1972

Freese, Barbara. Coal: A Human History. Penguin Books: London. 2003.

Korson, George.   Black Rock: Mining Folklore of the Pennsylvania Dutch. John   Hopkins Press: Baltimore. 1960.

Hatcher, John.   Before 1700: Towards the Age of Coal. Volume 1 of the History of The British Coal Industry.   Clarendon Press: Oxford.   1993

Richardson, Judith.   Possessions: The History and Uses of Haunting in the Hudson Valley.   Harvard University Press: Cambridge. 2003.

Sabol, John G. Jr. Ghost Excavator: Unearthing the Drama in the Mine Fields. Authorhouse: Bloomington, Indiana. 2007. Ghost Culture: Theories, Context, and Scientific Practice. Authorhouse: Bloomington, Indiana. 2007.

Schama, Simon. Landscape and Memory.   Alfred A. Knopf, Inc. : New York. 1995.

Skal, David J.   Death Makes a Holiday.   Bloomsburg: New York. 2002.

Young, James E.   The Texture of Memory, Holocaust, Memorials and Meaning. Yale University Press: New Haven.

About the Author

John Sabol has been participating in (and directing) scientific field investigations since 1969. He has conducted archaeological, anthropological, and paranormal research in England, Germany, Mexico, and throughout the United States. He is a former professor of inter-cultural studies, with 11 years of teaching experience at various universities in Mexico City, and in Guadalajara. His first book, Ghost Excavator, is a personal account of various paranormal experiences while growing-up in the anthracite coal region of Northeastern Pennsylvania in the 1960's. He is the author of three other books, Ghost Culture, Gettysburg Unearthed, and Battlefield Hauntscape. He has a M.A. in Cultural Anthropology (with a minor in Archaeology), and a B.A. in Sociology. He currently resides in Mahanoy City, Pennsylvania, centrally located in the southern fields of the anthracite coal region, with his daughter, Melissa. For more information about his books and paranormal investigations, please refer to his web sites: www.ghostexcavator.com and http://mysite. verizon.net/vzeoqapc/ghostexcavator.